From the Big Bang to You

Joey Raines

Published by Mirror Loop Publishing, 2025.

FROM THE BIG BANG TO YOU

First edition. December 1, 2025.

ISBN: 979-8232026035

Written by Joey Raines.

Table of Contents

For those who look up at the night sky, who wonder where we came from, and who try to understand our place in the universe.

To understand ourselves, we must understand where we came from.

Preface

Every book begins for a reason, and this one began with a question that would not leave me alone. I wanted to understand how everything started. Not in a spiritual sense, not in a symbolic sense, but in the literal scientific sense. What happened in the beginning. How did energy become matter. How did matter become stars. How did stars build the elements that would one day become oceans, planets, and living cells. How did those cells become conscious beings capable of looking back at the cosmos that produced them.

I had spent years hearing fragments of the scientific story. A piece from physics here. A piece from biology there. A diagram of the early universe. A lesson in evolution. A description of how stars live and die. Each piece was fascinating on its own, yet it felt incomplete. The full story was scattered across textbooks, documentaries, research papers, and conversations. I wanted to bring those pieces together in one place. I wanted a clear, continuous narrative that moved from the Big Bang to the formation of the first atoms, from the birth of stars to the rise of life, from early organisms to human consciousness. I wanted to read the entire arc from cosmic origins to human life in one steady line.

That book didn't exist in the form I needed, so I wrote it.

This book is not a spiritual argument, and it is not built on theology. It presents the scientific account alone, the story supported by observation, experiment, and evidence. It follows what the scientific community currently believes about the origin and development of physical reality. It walks through the steps that shaped the cosmos long before life emerged, and it explains how those same steps created the conditions that allow you to exist today.

At the same time, I hold a personal belief that does not interfere with the science presented here. I believe that God is the first scientist and the last scientist, the One who understood the laws before they existed and who set the stage for everything that followed. That belief belongs to me. The chapters in this book belong to science. They can stand together without conflict because they address different questions. One speaks to how. The other speaks to why.

My goal in writing this was simple. I wanted to give you a clear path through the entire scientific story. I wanted to make it readable, understandable, and meaningful without requiring a background in physics or biology. I wanted to show how one event leads to the next, how simple beginnings build toward complexity, and how the long chain from the first instant of time reaches all the way to you.

If this book helps you see your place in that chain, if it helps you understand the pathway from the Big Bang to human life, if it gives you a sense of connection to the history written into your atoms, then it has done its work.

Thank you for reading.

Introduction

Every question humanity has ever asked begins in the same place. It begins with curiosity. A simple thought crossing the mind. A moment when you look at the world and realize that everything you see has a history older than memory. The light from the Sun took eight minutes to reach you, traveling through the void at a speed nothing can exceed. The atoms in your hands were formed in the cores of ancient stars that burned for millions of years before exploding into space. The space around you expands every second, carrying distant galaxies away from one another at velocities that grow with distance. You are living inside a reality that has been unfolding for almost fourteen billion years, yet most of its story remains hidden from ordinary sight. The air you breathe, the ground beneath your feet, the thoughts moving through your mind right now, all of it traces back to a beginning so extreme that the laws of physics themselves were still taking shape.

This book was written to walk you through that story. Not as a textbook and not as a scientific manual, but as a narrative that begins at the simplest possible beginning and ends with the most complex thing the cosmos has ever produced. Awareness. Your awareness. The ability to ask where everything came from and why it evolved the way it did. The capacity to stand in the present moment and reconstruct the past through evidence, to peer backward across billions of years and see the pathway that led from formless energy to structured matter, from lifeless chemistry to living cells, from simple organisms to creatures who can contemplate their own existence.

Everything in these pages describes what science currently believes about the origin and development of physical reality. It follows the evidence collected through observation, experiment, and mathematical reasoning. Telescopes that capture light from galaxies so distant their photons have been traveling for billions of years. Particle accelerators that recreate conditions not seen since the first fractions of a second after the beginning. Microscopes that reveal the molecular machinery inside cells. Fossils that document the transformation of life across geological time. Genetic sequences that trace ancestry back through millions of generations. This book presents the account built from that evidence, the picture assembled piece by piece from physics, chemistry, biology, and cosmology. It does not present a spiritual view or a religious explanation. It presents the scientific account alone, without speaking for faith or replacing it. The questions of meaning, purpose, and what lies beyond the reach of observation remain open. Science describes how the physical world operates. It does not claim to answer why it exists or what significance that existence holds.

The chapters ahead follow the chain of events that shaped reality as we know it. They start in the first instant after the beginning, when the entire cosmos was smaller than a grain of dust and hotter than any furnace could imagine. In that moment, all the energy that would eventually become stars, planets, and living things existed in a state so compressed and so violent that matter as we recognize it could not yet form. They move through the creation of particles as the universe cooled, the birth of atoms when temperatures finally dropped enough for electrons to bond with nuclei, and the rise of stars that illuminated the darkness and began the long work of building complexity. They explore how those stars forged every element found in planets, oceans, and biological life, synthesizing carbon, oxygen, nitrogen, iron, and dozens of other elements through nuclear fusion in their cores and scattering them across space when they died. They follow the steps from molecules to cells, from simple organisms to creatures with nervous systems, and finally to minds capable of thought, memory, and the construction of knowledge. Each stage builds on what came before. Each transition opens new possibilities that didn't exist in earlier eras.

Along the way you will see how science understands the unfolding of reality. You will learn how matter organizes itself according to forces that operate across all scales, how gravity pulls gas into stars and shapes galaxies into spirals, how electromagnetic forces bind atoms into molecules and molecules into the structures of life. You will see how life adapts through evolution, how random mutations and natural selection sculpt organisms across generations, refining survival strategies and building complexity without intention or foresight. You will learn how intelligence emerges from the interactions of billions of neurons, how networks of cells generate perception, emotion, language, and the sensation of being someone experiencing the world. You will see how the laws of nature shape everything from the motion of galaxies to the chemistry inside your cells, how the same principles govern processes separated by incomprehensible differences in scale and time. And you will encounter the questions that science has not yet answered, mysteries that remind us how much remains unknown. What came before the beginning, if anything. Why the laws of physics have the particular forms they do. How consciousness arises from matter. Whether life exists elsewhere in the cosmos. These questions stand at the boundary of current knowledge, pointing toward what future discovery might reveal.

This book does not tell you what to believe. It gives you the scientific account of how the physical world formed and how life became possible. It presents the evidence and the explanations science has constructed from that evidence. The meaning you take from this belongs to you. Whether you see the cosmos as a natural process unfolding according to impersonal laws, a deeper design reflecting intention or purpose, or a mystery that still invites discovery and wonder, your interpretation has a place here. The facts of cosmology, evolution, and neuroscience describe what happened. They don't prescribe what that happening signifies. Science measures, predicts, and explains. It does not assign value or dictate philosophy. That work belongs to human reflection, to the conversations we have about what it means to exist in a universe that produced us through processes we are only beginning to understand.

You are part of this story whether you realize it or not. You are made from ancient matter, from atoms forged in stellar cores and scattered across space by supernovae, gathered together by gravity into clouds that collapsed to form the solar system, assembled into a planet where chemistry became biology. You are shaped by biological history, by evolutionary paths stretching back through millions of ancestors, each shaped by environments that tested survival and reproduction. You think with a brain carved by evolution over millions of years, an organ built from trial and error across countless generations, refined by the demands of social cooperation, language, memory, and abstract reasoning. And now, in this moment, you have the ability to look back across the full arc of existence and ask the question that inspired this book.

How did everything begin?

The chapters that follow will walk you through that answer, from the smallest scales of physics to the widest reaches of the cosmos, from the quantum realm where particles flicker in and out of existence to the observable universe stretching billions of light years in every direction. They will trace the development of complexity from simplicity, the emergence of life from chemistry, the rise of awareness from biological tissue. And by the time you reach the end, you may find that the story of the universe is also the story of you. Not metaphorically, but literally. You are what the cosmos has become after fourteen billion years of evolution. Your existence is the current expression of processes that began before planets existed, before stars formed, before matter itself had cooled enough to hold stable form. Understanding that history means understanding yourself in a way that no previous generation could access. It means seeing your place not just in human civilization or biological evolution, but in the complete narrative of physical reality.

Chapter One
The Big Bang

The story of the universe begins with a moment we cannot witness directly, yet its effects surround us in every direction. Every atom in your body, every star in the sky, every planet that circles a distant sun can trace its origins back to a single event. Scientists call this moment the Big Bang. It marks the beginning of the universe as we know it. Nothing we understand today existed before it. There was no space to expand into, no time to measure, no matter to arrange itself into stars or galaxies.

The Big Bang was not an explosion in space, as the name might suggest. It was something more fundamental and strange. It was the sudden appearance of space itself, filled with energy and possibility. There was no center to this event, no edge where the explosion occurred. Instead, every point in the universe began expanding away from every other point simultaneously. Space itself stretched like the surface of a balloon being inflated, except there was no inside or outside to this balloon. There was only the surface, and that surface was the entire universe. This is difficult to visualize because we are accustomed to thinking of explosions happening within existing space. The Big Bang created space rather than expanding into it.

In the first fraction of a second after the Big Bang, the universe expanded with unimaginable speed. This expansion was not like objects moving apart through a pre-existing void. It was space itself stretching, carrying everything within it. The distinction matters because it helps explain how galaxies can move away from us faster than the speed of light without violating the laws of physics. They are not racing through space at impossible speeds. Space itself is expanding between us and them.

The temperature at the earliest moments was unimaginably high, far beyond anything we can recreate in laboratories or observe in nature today. In the first trillionth of a second, the temperature exceeded ten trillion trillion degrees. Matter in any stable form couldn't exist under these conditions. Instead, the universe was a sea of energy, too hot for particles to hold together for more than fleeting instants. Even the distinction between energy and matter was blurred. The universe was governed by quantum effects at scales we can barely comprehend.

Time began in this moment. Before the Big Bang, the word "before" has no meaning. Time itself came into existence along with space and energy. With time came the possibility of change, of events following one another in sequence. The laws of nature started to take shape as the universe cooled. Gravity, electromagnetism, and the nuclear forces emerged, separating from a unified state we still don't fully understand. These forces would later control the formation of atoms, stars, planets, and life. In the earliest moments, however, they may have been unified into a single fundamental force. As the universe expanded and cooled, this unity broke apart into the distinct forces we observe today.

Gravity is the weakest of these forces, yet it operates over infinite distances and shapes the structure of the cosmos. Electromagnetism governs the behavior of charged particles and makes chemistry possible. The strong nuclear force binds protons and neutrons together in atomic nuclei, overcoming the electromagnetic repulsion between positively charged protons. The weak nuclear force governs certain types of radioactive decay and played a crucial role in the early universe. Each force has its own character, its own domain of influence. Together, they created the conditions necessary for complexity to emerge from simplicity.

As the universe continued to expand, it cooled just enough for energy to transform into particles. This transformation follows Einstein's famous equation linking energy and mass. When conditions are right, pure energy can materialize as matter. The simplest ingredients of matter appeared first. Quarks and electrons formed out of the intense energy, flickering into existence and occasionally annihilating with their antimatter counterparts. Quarks are the building blocks of protons and neutrons, though they never exist alone in nature. They are always bound together in groups.

In those first microseconds, quarks briefly paired into protons and neutrons before being torn apart by the heat. The temperature was still too high for these composite particles to remain stable for long. But as the universe continued expanding and cooling, a critical threshold was crossed. Around one millionth of a second after the Big Bang, the temperature dropped below a trillion degrees. At this point, quarks could bind permanently into protons and neutrons. This period was chaotic, violent, and brief. Yet it laid the foundation for everything that would follow. Elementary particles were the first building blocks, destined to join together in countless combinations over billions of years.

A few minutes after the Big Bang, the universe cooled enough for protons and neutrons to join into small atomic nuclei. This process, called nucleosynthesis, created the first elements. The temperature had fallen to about a billion degrees, similar to the core of the hottest stars. At this temperature, nuclear fusion could occur. Protons combined with neutrons to form deuterium, a heavy form of hydrogen. Deuterium nuclei then fused together to create helium. Some of these reactions produced lithium, though in much smaller amounts.

Most of the universe became hydrogen, the lightest and simplest element, consisting of a single proton. About one quarter of the ordinary matter became helium, with two protons and two neutrons in each nucleus. Only a trace became lithium. These proportions were determined by the balance between the expansion rate and the nuclear reaction rates. If the universe had expanded more slowly, more helium would have formed. If it had expanded faster, less would have formed. The fact that we observe roughly twenty five percent helium throughout the universe, in stars both young and ancient, is powerful evidence that these calculations correctly describe the early universe.

These simple nuclei carried the signature of the early universe within their structure. They would one day form stars, planets, oceans, and living beings. But for now they drifted in a young universe still too hot for entire atoms to form. A nucleus is not yet an atom. It is only the dense core, lacking the surrounding cloud of electrons that defines an atom's chemical behavior. The universe needed to cool further before atoms could assemble.

For hundreds of thousands of years after nucleosynthesis, particles moved through space without attaching to one another in stable configurations. Free electrons roamed through the cosmos, not yet captured by atomic nuclei. Photons, the particles of light, scattered in all directions, unable to travel far before being absorbed or deflected by these free electrons. The universe was filled with light, but none of it could escape the constant interactions with matter. It was a glowing fog, dense and hot, where matter and radiation blended together in a state of equilibrium.

Imagine trying to see through a thick cloud. Light enters but scatters from water droplets, making it impossible to see clearly even a short distance ahead. The early universe was similar, except the scattering came from electrons rather than water droplets. This meant the universe was opaque despite being filled with intense radiation. No structures could be seen at a distance. Everything was veiled in this primordial fog.

During this time, small irregularities in density appeared. Some regions were slightly more dense than others. These differences were tiny, perhaps only one part in one hundred thousand, but they were crucial. Gravity amplifies small differences over time. Denser regions pulled in more matter, becoming denser still. These slight variations would later become the seeds of galaxies and clusters, shaping the large-scale structure of the universe. Without these initial irregularities, the universe might have remained forever smooth and featureless. Stars and galaxies would never have formed.

Eventually, about three hundred eighty thousand years after the Big Bang, the universe cooled enough for electrons to attach to atomic nuclei. The temperature had fallen to about three thousand degrees, cool enough for atoms to remain stable. At that point, atoms formed for the first time. Electrons settled into orbit around nuclei, creating neutral hydrogen and helium atoms. This moment is called recombination, though the name is misleading because the electrons and nuclei had never been combined before.

When this happened, light could suddenly travel freely through space without being scattered. With no free electrons left to block their path, photons streamed through the universe unimpeded. The fog cleared, and the universe became transparent. This was a transformation as dramatic as a sunrise, except it happened everywhere at once. The light released in that moment still exists today. It has been traveling through space for nearly fourteen billion years, stretching and cooling as the universe expanded. It fills the entire cosmos as a faint glow called the cosmic microwave background.

This radiation is not visible to the naked eye. It has been stretched by the expansion of space from visible light into microwave wavelengths. But we can detect it with specialized instruments. It arrives from every direction with remarkable uniformity, showing the same temperature to within one part in one hundred thousand no matter where we look. This uniformity tells us the early universe was extraordinarily smooth and homogeneous. The tiny variations in this radiation map the density fluctuations that would grow into galaxies. The cosmic microwave background is the oldest light we can detect, a silent message from the early universe that confirms our understanding of how everything began.

With atoms now stable and light able to travel across space, the universe entered a long period of gradual change. The dramatic events of the first few minutes were over. What followed was a slow gathering of matter under gravity's patient pull. Clouds of hydrogen and helium drifted through space, slowly pulling together. Gravity drew atoms toward regions that were slightly denser than average. Over millions of years, these clouds grew denser and hotter at their cores until the pressure and temperature became sufficient for nuclear fusion to begin.

The first stars ignited perhaps one hundred to two hundred million years after the Big Bang. They were made entirely of hydrogen and helium, the only elements that existed in significant quantities. These stars were likely more massive than most stars today, burning hotter and dying faster. When they exhausted their fuel, some exploded as supernovae, scattering newly forged elements into space. Carbon, oxygen, nitrogen, iron, and dozens of other elements were created in the cores of these stars and in their explosive deaths. From those stars came not only heat and light but the heavier elements that would someday form planets, oceans, mountains, and the molecules of life.

The Big Bang set these events in motion. Without it, there would be no stars, no galaxies, no Earth, and no beings capable of asking how it all began. Every stage of cosmic evolution depended on the conditions established in those first moments. The right balance of forces, the right mixture of matter and energy, the right rate of expansion. If any of these had been significantly different, the universe might have collapsed back on itself immediately, or expanded so rapidly that matter could never coalesce into structures, or failed to produce the elements necessary for complexity.

The Big Bang does not explain why the universe exists. That question remains beyond the reach of science, at least for now. But the Big Bang explains how the universe started. It describes the first moments of time and the way space unfolded from an initial state of unimaginable density and temperature. It shows how simple particles arose from energy, and how the early universe prepared the conditions needed for all future complexity. Every chapter of cosmic history depends on this beginning, from the birth of elementary particles to the formation of atoms, from the ignition of the first stars to the eventual emergence of life on at least one small planet orbiting an ordinary star.

The universe didn't begin in silence or darkness. It began in heat, expansion, and transformation. It began with energy becoming matter, with space itself coming into existence, with time starting its forward march. In a sense, everything that has happened since has been an elaborate consequence of those first moments. Stars, galaxies, planets, and life are all footnotes to the story that began with the Big Bang.

Chapter Two
Time

Time is so familiar to us that we rarely question what it actually is. We wake up, we move through the day, and we watch moments pass in a steady stream, as if time is simply the background rhythm of life. We measure it with clocks, divide it into hours and seconds, and feel it pressing forward with each heartbeat. Yet despite this intimacy, time is more mysterious than anything we see around us. It is not an object we can touch or examine. It is not a substance that flows like water. It is not a force that pushes or pulls. Time is the dimension along which events unfold, the framework that gives order to cause and effect. It is what allows one moment to become the next, what transforms possibility into memory. Without time, nothing could change, nothing could move, nothing could begin, and nothing could end. The universe needed time before it could build even the smallest particle, and understanding time is the first step toward understanding how everything else came to be.

In the earliest moment of the universe, time itself began. That moment is not like anything we experience in daily life. It was not a place you could visit, and it was not a point sitting inside a larger space waiting to be discovered. It was the beginning of the entire fabric that would support both time and space, the origin of the stage and the play together. Before that moment, there was no earlier instant, because there was no flow of time at all. The very concept of "before" requires time to exist first, which means the question dissolves into meaninglessness at the boundary of the universe's birth. Asking what came before the beginning of time is like asking what is north of the North Pole. The idea appears to make sense at first, until you realize that it asks for something the universe does not provide. There is no direction that takes you beyond the starting point, because directions themselves only exist after the framework is laid.

Once time began, it moved in one direction only. It flowed forward, carrying the universe from simplicity toward complexity, from the initial burst of energy toward the structures we see today. In our experience, we call this the arrow of time, and it points unwaveringly from past to future. Cups fall and break, but broken cups don't leap up and reassemble themselves. We grow older and never grow younger. Heat moves from hot objects to cold ones, never traveling the other direction without external effort. These simple experiences reflect a deep truth about the universe. Time has a direction because the universe began in a state of extreme order, all its energy compressed and organized in ways that could never happen by chance, and it has been drifting toward disorder ever since. That increase in disorder is what physicists call entropy, a measure of how spread out and randomized energy becomes over time. Entropy is the reason the future always looks different from the past. It is the reason eggs scramble but don't unscramble, the reason perfume spreads through a room but never gathers itself back into the bottle. The arrow of time emerges from this relentless slide toward greater disorder, giving the universe a direction even though the fundamental laws of physics themselves show no preference for forward or backward.

Time is also connected to space in a way that surprised every scientist who first learned it. For thousands of years, people believed time was absolute, a universal clock ticking at the same rate everywhere in the cosmos, independent of anything happening within it. We think of space as the place where things exist and time as the sequence of events, treating them as separate realities. But they are not separate. The two are woven together into a single four-dimensional fabric. Together they form what scientists call spacetime, a unified structure where the three dimensions of space and the one dimension of time blend into something inseparable. The flow of time is affected by the presence of matter and energy, responding to the curvature they create in spacetime itself. Wherever there is mass, time passes slightly slower. This effect is small in everyday life, too tiny to notice as we walk through our days, but on cosmic scales it becomes enormous and undeniable. Clock time on Earth runs a little more slowly than time in empty space because Earth's gravity bends spacetime around it, creating a subtle depression in the fabric that slows the ticking of every clock and the beating of every heart. Time near a black hole slows even further, so much that an observer far away would watch events near the hole appear to creep along in slow motion, nearly frozen. An astronaut falling toward a black hole would experience time normally from their own perspective, but to someone watching from a safe distance, that astronaut would seem to take forever to cross the boundary, their movements grinding down to an imperceptible crawl. Time is not a constant background ticking uniformly across the universe. It stretches and curves with the universe itself, responding to the distribution of mass and energy like a flexible surface responds to weight.

The speed at which time passes is also tied to motion through space. If you move very fast compared to someone else, time for you will pass more slowly relative to them. This is not science fiction or philosophical speculation. It is a measurable effect that has been confirmed again and again through careful experiments. Astronauts on the International Space Station, moving at roughly 17,500 miles per hour, age a tiny bit slower than people on Earth due to their speed. When they return after months in orbit, they have experienced a few milliseconds less time than those who remained on the ground. The difference is extremely small, but it is real and measurable with precise atomic clocks. Time is flexible in ways that defy our intuition. It changes depending on motion and gravity, which means our experience of time, sitting relatively still on the surface of a planet in a quiet corner of the galaxy, is only one version of how time behaves across the universe. Two people can experience different amounts of time passing between the same two events, and both perspectives are equally valid. There is no single universal "now" that everyone shares.

Although time shapes every moment of our lives and gives meaning to our memories and plans, the universe does not treat it with the same emotion we do. The universe does not feel time. It does not experience waiting or urgency. It does not rush or linger. It simply unfolds according to the laws that govern it, indifferent to whether a process takes a second or a billion years. Stars form from collapsing clouds of gas, burn through their fuel, and die over billions of years, timescales so vast they render human lifetimes nearly invisible. Galaxies drift apart over immense stretches of time as space itself expands between them. Planets cool from molten spheres to solid worlds, oceans rise and fall, mountains erode grain by grain, and species come and go, all following the steady rhythm of a cosmic clock that does not hurry and does not pause. Time allows the universe to create things slowly, step by step, following the laws of nature with patient consistency. The universe has no deadlines. It moves at the pace the laws of physics allow, transforming energy and matter through processes that can take mere moments or ages beyond comprehension.

Without time, particles could never evolve into atoms, because evolution requires a sequence of events, a progression from one state to another. Atoms couldn't gather into stars, because gravity needs time to pull matter together across the void. Stars couldn't forge the heavier elements needed for planets and life, because nuclear fusion is a process that requires billions of years of sustained pressure and heat. Life itself requires time to grow, change, adapt, and become more complex, building from simple chemistry to organisms capable of thought and wonder. Everything that exists is shaped by the fact that time flows forward and that this flow creates the possibility of change, of cause leading to effect, of simple structures giving rise to complex ones. Time is not simply something the universe contains, like stars or galaxies scattered through space. It is something the universe uses to build itself, the dimension that allows potential to become actual, the framework that transforms the possible into the real.

We live inside time and move through it without effort, carried forward whether we choose to be or not, but understanding it reveals something important about the nature of reality. Time is not separate from the story of the universe. It is not an afterthought or a detail that could have been otherwise. It is the beginning of the story, the very first word that makes all other words possible. It is the foundation that every chapter depends on, the ground on which every structure stands. Without time, there would be no universe to speak of, no change to measure, no events to connect. As we continue this journey through the architecture of reality, we will move from time into the space that surrounds us, the stage on which the universe expresses itself, and the structure that allows everything else to exist. But we should remember that time came first, and everything that follows emerges only because time opened the door.

Chapter Three
Space

Space seems empty when we look at it, a vast darkness that stretches in every direction and feels impossible to imagine. Yet space is not nothing. It is not a blank container waiting to be filled. Space is part of the fabric of the universe, as real and fundamental as matter itself. It has structure, it has shape, and it responds to the presence of matter and energy in ways that determine how the universe behaves. Everything that exists, from galaxies to atoms to the cells in your body, exists within the arena of space. To understand how the universe builds anything, we must first understand the stage on which all of creation takes place.

In the earliest moment of the universe, space didn't exist. There was no void, no emptiness, no distance waiting to be measured. When the universe began, space unfolded along with time. They came into existence together as part of a single continuous fabric that physicists call spacetime. That fabric stretches and bends. It can expand. It can curve. It can ripple like the surface of a pond when a stone falls through it. It can even twist under the right conditions, spiraling in ways that would seem impossible if we thought of space as mere emptiness. This is very different from the ordinary sense we have of distance. When you walk across a room, it feels as if you are moving through something empty, crossing a neutral gap between one place and another. In reality, you are moving across a piece of spacetime that is gently curved by the mass of Earth beneath you. The floor under your feet is not just supported by wood or concrete. It is held in place by the warping of space itself.

The idea that space and time form a unified fabric was one of the most profound insights in the history of science. Before the early twentieth century, scientists treated space as an unchanging backdrop, a static grid through which objects moved. Time was thought to flow at the same rate for everyone, regardless of their motion or location. But observations showed that this picture was incomplete. Light behaves in ways that cannot be explained if space and time are separate. The speed of light remains constant no matter how fast you are moving, which should be impossible if time flows the same way for everyone. The resolution came from understanding that space and time are woven together. When you move through space, you also move through time. When you change your speed, the rate at which time passes for you changes as well. Space and time are not independent dimensions. They are aspects of a single structure that can be measured and described but cannot be separated.

Gravity is not a force that pulls objects together the way a magnet pulls metal. It is the curvature of space caused by mass. When a planet or star sits in space, the space around it bends. The more massive the object, the more dramatic the bending becomes. Other objects follow the curves created by that bending, traveling along what mathematicians call geodesics. A geodesic is the shortest path between two points in curved space, much like the arc of a great circle on the surface of a sphere. We experience this as gravity. When Earth moves around the Sun, it is not being dragged by a force reaching across the emptiness between them. It is following the shape of space created by the Sun's mass, moving in an orbit because that orbit is the natural path through curved spacetime. When you fall, you are not being yanked downward by an invisible hand. You are moving along the curved path of spacetime around Earth, following the geometry that Earth's mass has created. Space is not passive. It shapes the motion of everything inside it, from the smallest asteroid to the largest galaxy.

This understanding transforms how we think about weight and motion. When you stand on the ground, you are not at rest. You are constantly accelerating, trying to follow the curved path of spacetime that would take you toward the center of Earth. The ground beneath you pushes back, preventing you from following that natural path. This is why you feel weight. The sensation you interpret as gravity pulling you down is actually the feeling of being held up, pushed away from the path you would naturally follow through spacetime. An astronaut in orbit experiences weightlessness not because gravity has disappeared but because nothing prevents them from following their natural geodesic path. They are in constant free fall, moving through curved spacetime without obstruction.

Space also expands. This is one of the most surprising discoveries in science, and one of the most difficult to grasp intuitively. The universe is not expanding into anything. There is no outer boundary where space meets something else. Space itself is growing. The distance between galaxies increases over time because new space is forming between them, stretching the fabric of the cosmos. This expansion has been happening since the beginning, and it continues today. It is not something we notice in daily life. Distances on Earth remain stable because gravity holds matter together. The gravitational binding between atoms, between molecules, between the particles in your body is strong enough to resist the expansion completely. Even the distance between Earth and the Sun stays constant. The expansion only becomes noticeable across vast cosmic distances where gravity is too weak to counteract it. The farther away a galaxy is, the faster it appears to move, not because it is speeding through space but because the space between us and that galaxy is stretching.

We discovered this expansion by observing light. When a galaxy moves away from us, the light it emits stretches along with the expanding space. This stretching shifts the light toward the red end of the spectrum, a phenomenon called redshift. The more distant a galaxy is, the greater its redshift, revealing that expansion increases with distance. This pattern was first observed in the 1920s and has been confirmed by decades of careful measurement. It tells us that the universe was once much smaller and denser than it is today. If we trace the expansion backward in time, we find that all of space was compressed into an unimaginably small region. This is the beginning we call the Big Bang, the moment when space and time themselves came into existence.

Even though space is enormous, it is not uniform. It contains invisible structures shaped by something we cannot see. Scientists call this unseen ingredient dark matter. Dark matter does not emit light, reflect it, or absorb it. It passes through ordinary matter almost without interaction, moving through planets and stars as if they were not there. Yet it has gravity. It curves space just as ordinary matter does, and its gravity is strong. There is far more dark matter in the universe than ordinary matter. For every atom in a star, in a planet, or in your body, there are roughly five times as much mass in dark matter surrounding it. This dark matter creates the scaffolding that galaxies use to form. It clumps together in vast halos that pull ordinary matter inward, providing the gravitational framework for stars to gather and galaxies to take shape.

Without dark matter, the universe would not look the way it does. Galaxies would not cluster into the groups and superclusters we observe. Stars would not gather into spirals and ellipses with the rotation speeds we measure. When astronomers observe how fast stars orbit at the edges of galaxies, they find that those stars move much faster than they should if only visible matter were present. The gravity from the stars and gas we can see is not strong enough to hold those outer stars in orbit. They should fly apart, scattering into space. Yet they remain bound, held in place by the gravity of dark matter that extends far beyond the visible edge of the galaxy. The cosmic web that fills the universe, the great filaments of galaxies stretching across billions of light years with vast voids between them, would not exist without dark matter. Space is held together by matter we cannot detect directly, but whose influence we can measure through the behavior of galaxies, the bending of light around massive clusters, and the patterns imprinted in the radiation left over from the early universe.

On even larger scales, space contains another mystery. It is influenced by something that behaves like an outward pressure, a kind of antigravity that does not pull matter together but pushes space apart. This pressure causes the expansion of the universe to accelerate. Scientists call this dark energy. It has no clear source and no known structure. It does not clump like matter. It does not weaken as space expands. Instead, it appears to be a property built into the nature of spacetime itself, a constant energy density that fills every cubic centimeter of space equally. Its effect grows stronger as the universe grows larger simply because there is more space for it to occupy. As expansion continues, the volume of space increases, and with it the total amount of dark energy. This creates a feedback loop. More dark energy leads to faster expansion, which creates more space, which contains more dark energy, which accelerates expansion further.

Dark energy may determine the future shape of the cosmos, guiding how far space will stretch and what kind of universe will exist trillions of years from now. If dark energy remains constant, the universe will expand forever, growing colder and more diffuse until galaxies are separated by distances so vast that light from one will never reach another. If dark energy strengthens over time, expansion could accelerate without limit, eventually tearing apart galaxies, stars, planets, and even atoms in what cosmologists call the Big Rip. If dark energy weakens, gravity might eventually overcome expansion, pulling everything back together into a Big Crunch. We don't yet know which future awaits us, but we know that dark energy will play a decisive role.

Space is not empty. It is filled with fields, forces, particles, and invisible influences that give it texture and shape. Even the vacuum, the region that seems perfectly empty when we remove all matter and radiation, is alive with activity. Quantum mechanics tells us that empty space is not truly empty. The vacuum seethes with energy. Tiny particles flicker into existence and vanish again in moments too brief to measure directly. These are called virtual particles, and they appear because of the uncertainty principle that governs the quantum world. For an infinitesimally short time, energy can be borrowed from the vacuum to create a particle and its antiparticle partner. These pairs emerge, exist for a fraction of a second, and then annihilate each other, returning the borrowed energy. We cannot see these particles directly, but we can measure their effects. They create a subtle pressure that has been detected in laboratory experiments. They affect the energy levels of atoms in ways that have been measured to extraordinary precision. These fluctuations are a sign that space has energy built into it, that the vacuum itself is not nothing but a dynamic, restless entity.

Space can stretch and bend because it contains the ability to change. It is not a rigid stage but a responsive medium that reacts to the presence of matter and energy. It is not a silent void. It is an active participant in the universe, shaping the paths of light, the orbits of planets, the clustering of galaxies. When massive objects accelerate or collide, they send ripples through spacetime itself. These ripples travel at the speed of light, carrying energy away from their source. We call them gravitational waves. They were predicted a century ago, but it took decades of technological advancement before we could build instruments sensitive enough to detect them. When two black holes spiral together and merge, they shake the fabric of space violently enough that we can measure the disturbance here on Earth, even though it occurred billions of light years away. The detection of these waves confirmed that space is not merely a concept or a mathematical abstraction. It is a physical entity that can vibrate, carry energy, and transmit information across the cosmos.

Everything we will explore in the following chapters depends on the nature of space. Without space to expand, the universe couldn't grow from its initial state. Without space to curve, gravity couldn't shape stars or planets or create the conditions for matter to gather into structures. Without the structure of space, light would have no path to travel, no medium through which electromagnetic waves could propagate. Life depends on the stability of distances, which allows atoms to bond and molecules to form. Life depends on the behavior of energy, which flows through space according to predictable laws. Life depends on the predictable laws of motion, which arise from the geometry of spacetime. All of these arise from the nature of space itself.

As we move forward, we will leave the fabric of space and step into the earliest events that shaped the universe. The beginning of everything marked the moment when space and time opened and the foundations for matter were set in place. From that moment, the universe began building toward the first particles and eventually toward every structure that would follow. Understanding space is the first key to understanding the story of creation in all its forms. It is the foundation upon which everything else is built, the stage that makes all other acts possible. Without space, there would be no universe to explore, no matter to study, no observers to ask questions. Space is where the story begins.

Chapter Four

The Beginning Before Everything

The universe has a beginning, and that beginning is unlike anything we experience or can fully imagine. It was not a moment where something exploded into empty space. There was no space to explode into. There was no outside, no before, no surrounding silence. The beginning was the moment when space itself appeared, along with time, energy, and the laws of nature. It was the start of everything that would ever exist.

To understand what this means, we must think differently about the nature of beginnings. When we imagine an explosion, we picture something bursting outward from a central point into an existing volume. But the beginning of the universe was not like that. Space itself came into existence, and as it expanded, it carried everything with it. There is no edge to this expansion, no boundary where the universe stops. Every point in space is expanding away from every other point, not because matter is flying through space, but because space itself is growing. To understand the story of the universe, we must accept that the beginning is not a place we can travel to or an event we can witness from the outside. It is the moment when the stage and the actors appeared at the same time.

In the first instant, the universe was incredibly small, unimaginably dense, and hotter than any environment that exists today. When we say small, we mean smaller than an atom, smaller than the nucleus of an atom, smaller even than the smallest particles we know. All the energy that would ever become matter was compressed into a space so tiny that the laws of physics as we understand them begin to break down when we try to describe it. The temperature was beyond comprehension, trillions of degrees, far hotter than the center of any star. At such temperatures, matter as we know it cannot exist. There were no atoms, no particles, and no structures of any kind. The universe was a hot, dense, expanding sea of pure energy.

It was not chaotic. It was simply uniform and featureless, a cosmic blur with no variation. The energy was spread evenly, and the conditions were the same everywhere. This uniformity is one of the most remarkable features of the early universe. If you could somehow observe it from one location, it would look the same as from any other. Yet within this near perfect smoothness, quantum effects produced tiny fluctuations, irregularities so small they would have been invisible to any observer. These variations, almost too small to notice, would eventually guide the formation of galaxies, stars, and planets. They were the seeds of all structure.

As the universe expanded, time moved forward, and the temperature dropped. This cooling was not like the cooling of a hot object in a room. There was no surrounding cold space to draw away the heat. Instead, the expansion of space itself caused the cooling. As space grew, the energy within it became less concentrated, and the temperature fell. This cooling allowed energy to transform into the very first building blocks of matter. Here we encounter one of the most profound relationships in physics. Energy and matter are two forms of the same thing, connected by the equation that states energy equals mass multiplied by the speed of light squared. In the intense heat of the early universe, energy could spontaneously become particles, and particles could dissolve back into energy.

The earliest particles formed in the first fraction of a second, in a time span so brief that no clock we could build would measure it. They flickered in and out of existence as the intense conditions allowed for brief moments of stability. These particles were not yet stable enough to create the world we know. The universe had to cool further. During these early moments, the laws of nature behaved in ways we still don't fully understand, but one thing is clear. The basic rules that govern everything today were beginning to appear. The forces that hold atoms together, the forces that power stars, and the force of gravity itself all emerged from the conditions of this earliest time.

The early universe was filled with radiation, and that radiation still exists today, stretched into cold microwaves by billions of years of expansion. Scientists call this the cosmic microwave background. It is the faint afterglow of the beginning, a whisper of the moment when the universe was young. We can detect this radiation from every direction in space, and it has nearly the same temperature everywhere we look, just a few degrees above absolute zero. This tells us something remarkable. The early universe truly was smooth and uniform.

Yet when we examine this background radiation with great precision, we find it is not perfectly smooth. There are tiny temperature differences, variations of only a few parts in a hundred thousand. These small imbalances tell the story of how matter would eventually gather. In regions that were slightly denser, gravity had a stronger pull, and over time, matter collected there. In regions that were slightly less dense, matter drifted away. Those small imbalances were essential. Without them, the universe would not have created anything. Gravity would have had nothing to work with, no slight advantages to amplify. There would be no stars, no planets, and no life. The cosmic background shows us both the remarkable uniformity of the beginning and the subtle variations that made everything possible.

Although the beginning feels mysterious, science gives us a clear picture of how the universe behaved in its earliest stages. We can trace its history back to when it was less than a second old, and we understand the processes that occurred during those moments with remarkable precision. We know how fast it expanded. We know how its temperature changed. We know how energy transformed into matter.

What we don't know is why it happened. The laws of nature allow a universe to appear, and the conditions of the beginning set everything in motion. Whether this event was part of a larger process, whether other universes exist beyond our ability to observe, or whether this was a unique moment, is still unknown. What matters for our story is that the universe had a beginning, and its earliest moments shaped everything that followed.

Within the first three minutes of the beginning, the universe had cooled enough for simple atomic nuclei to form. The temperature dropped below a billion degrees, still incredibly hot, but cool enough for protons and neutrons to stick together. These nuclei were mostly hydrogen, which is simply a single proton, and helium, which contains two protons and two neutrons bound together. The proportions were set by the temperature and density of the universe at that time. For every twelve hydrogen nuclei, roughly one helium nucleus formed. This ratio would remain essentially unchanged for billions of years. These nuclei floated in the expanding sea of radiation, surrounded by free electrons that were still too energetic to be captured. The universe had to cool further, far further, before electrons could settle into orbit around nuclei and complete the first atoms. That process would take hundreds of thousands of years. The journey from pure energy to stable matter had begun, and each step carried the universe closer to a world capable of structure.

This beginning was not an explosion that sent debris flying outward into an empty void. It was the expansion of space itself, carrying everything with it. Every galaxy, every particle, every bit of radiation moves away from every other because the space between them grows. That expansion continues today. The universe is larger now than it has ever been, and it grows every second. The most distant galaxies we can see are racing away from us at tremendous speeds, not because they are moving through space, but because space itself is expanding.

FROM THE BIG BANG TO YOU

Everything that exists, including you, exists because the universe had those first moments where the rules of physics allowed energy to cool, matter to form, and the foundations of creation to appear.

The story of how the universe began is the story of how possibility entered existence. Before the beginning, there were no possibilities, no potential, and no future. The early moments created the conditions needed for particles, then atoms, then stars, and eventually life. The path from pure energy to the complex structures we see today required each step to unfold in sequence. First, the cooling had to allow particles to form. Then, the particles had to combine into nuclei. Then, after a long wait, those nuclei had to capture electrons and become atoms. Only then could gravity begin its slow work of gathering matter into the first stars. As we move forward, we will see how the earliest particles settled into stable forms and became the building blocks of everything that would follow. The beginning is where the journey starts, but it is only the first step in the long path toward a universe filled with light, matter, and life.

Chapter Five
Quantum Behavior

The universe is built from particles so small that we cannot see them, yet their behavior shapes every moment of reality. These particles don't act like the objects we experience in daily life. A ball rolls, a leaf falls, a river flows, and each movement follows rules that feel predictable and familiar. We can watch a thrown stone arc through the air and know exactly where it will land. We can measure the speed of a falling object and predict its path with simple equations. But at the smallest scales, the rules change completely. The universe behaves in ways that seem impossible until we understand the nature of quantum behavior. This microscopic world is not chaotic. It is governed by precise laws, but those laws reveal a universe far stranger and more elegant than the one we see with our eyes. The transition from the predictable world of everyday experience to the quantum realm happens at scales smaller than atoms, where particles measure less than a billionth of a meter across. At this boundary, certainty gives way to probability, and the comfortable logic of cause and effect transforms into something more subtle and profound.

In the quantum world, particles behave like both particles and waves. This is not a metaphor or a convenient analogy used by physicists to simplify a complex idea. It is a fact of nature, confirmed by countless experiments over the past century. A single electron can act like a tiny point in space, striking a detector at one specific location with measurable impact. But it can also spread out like a wave and pass through two openings at once, creating an interference pattern that only waves can produce. When we fire electrons one at a time toward a barrier with two slits, each electron passes through as if it traveled through both openings simultaneously, interfering with itself before arriving at the detector. The pattern that emerges over time is not the pattern of particles following straight paths. It is the pattern of waves overlapping, with bright regions where the waves reinforce each other and dark regions where they cancel out. This happens even when we send the electrons through one at a time, which means each individual electron must somehow interact with both paths. The electron exists in multiple possible states until something interacts with it, at which point it settles into one outcome. This behavior does not mean the particle is confused or indecisive. It means the particle genuinely exists in a cloud of possibilities that collapse into a definite state when measured. This idea is called superposition, and it is one of the most remarkable features of quantum physics. Before measurement, the electron does not secretly occupy one position while we remain ignorant of it. The electron truly exists in many states at once, spread across space like a probability cloud. Only when we observe it does reality choose one outcome from all the possibilities.

The quantum world also displays a level of connection that seems impossible from our everyday perspective. Two particles that interact can become linked in a way that causes their properties to match even when they are separated by great distances. Imagine two electrons that collide and then fly apart in opposite directions. In the quantum world, these electrons can become entangled, meaning their properties remain correlated no matter how far they travel from each other. If we measure the spin of one electron and find it pointing up, we instantly know the other electron will have its spin pointing down, even if that electron has traveled to the other side of the galaxy. When one particle is measured, the other responds instantly, as if they share information outside the normal limits of space. This connection is called entanglement, and Albert Einstein famously called it "spooky action at a distance" because it troubled him deeply. It does not allow communication faster than light, because the outcome of each measurement appears random to the person making it. We cannot control what result we get, so we cannot send a message by manipulating one particle and having the other respond in a predetermined way. But entanglement does reveal that the universe is more unified than it appears. Entangled particles behave like parts of a single system, even when they are far apart. The separation in space does not break the quantum connection. They remain correlated in a way that has no parallel in our everyday experience, as if distance itself becomes meaningless at the quantum level. This phenomenon has been verified in laboratories countless times, with particles separated by kilometers still showing perfect correlation when measured.

Another important idea in quantum behavior is uncertainty. At the smallest scales, certain pairs of properties cannot be known at the same time with perfect accuracy. The more accurately we know a particle's position, the less accurately we can know its momentum. The more precisely we pin down where an electron is located, the less we can say about how fast it is moving or in what direction. This relationship is not a flaw in our measuring tools or a limitation of current technology. It is a fundamental property of the universe, discovered by Werner Heisenberg in the 1920s. Particles don't have exact values for both position and momentum until they need to interact with something. They exist in a range of possibilities, and this uncertainty is woven into the structure of reality. The act of measuring one property inherently disturbs the other, not because our instruments are clumsy, but because the particle itself does not possess both properties simultaneously with infinite precision. This creates a natural limit to how much information we can extract from the quantum world. The universe guards certain secrets not out of mystery, but out of necessity. If particles could have perfectly defined positions and momenta at the same time, atoms would collapse. Electrons would spiral into the nucleus, matter would become unstable, and the universe as we know it couldn't exist.

These strange behaviors might feel disconnected from daily life, but they control everything around us. The structure of atoms, the stability of matter, the glow of light, and the flow of electricity all depend on quantum laws. Without quantum mechanics, we cannot explain why atoms exist at all. In classical physics, an electron orbiting a nucleus should radiate energy constantly and spiral inward in a fraction of a second, causing every atom to collapse. But quantum mechanics saves atoms from this fate. Atoms hold together because electrons form stable patterns around atomic nuclei, patterns that can only exist because electrons behave like waves. These patterns are called orbitals, and they represent the allowed states where electrons can exist without losing energy. An electron in an atom does not orbit like a planet around a star. It occupies a probability cloud, a region of space where it is likely to be found, and this cloud takes specific shapes determined by quantum laws. The colors we see in the world emerge because electrons jump between energy levels inside atoms, releasing light in precise amounts called photons. A neon sign glows red because electrons in neon atoms drop from higher energy states to lower ones, releasing photons of a specific wavelength that our eyes perceive as red. The exact color of light from any substance reveals the quantum structure of its atoms. Even the devices we use every day, including computers, phones, and the very tools scientists use to study the universe, depend on quantum effects. The semiconductors in computer chips work because electrons tunnel through barriers they shouldn't classically be able to cross, a purely quantum phenomenon. Lasers produce their concentrated beams of light through stimulated emission, another quantum process. The magnetic hard drives that store information rely on the quantum property of electron spin. Our modern technological world is built on quantum foundations, even though most people never think about the quantum behavior underlying their phones and laptops.

Quantum behavior also shapes the early universe in ways that ripple through billions of years to reach us today. In the first moments after the universe began, quantum fluctuations produced tiny variations in energy across space. These variations were incredibly small, differences in density of only one part in one hundred thousand. Yet as the universe expanded, they stretched into large-scale patterns that determined where matter would gather. The quantum uncertainty that exists at the smallest scales was amplified by cosmic expansion until it affected regions millions of light years across. Galaxies formed in regions where the energy was slightly higher, where matter was slightly denser than average. Gravity pulled this matter together over billions of years, creating the galaxies, stars, and planets we see today. Meanwhile, emptier regions became vast cosmic voids, enormous bubbles of nearly empty space separating clusters of galaxies. The structure of the universe today is a magnified reflection of quantum activity from its earliest moments. When we look at the cosmic microwave background radiation, the faint glow left over from the early universe, we see a pattern of hot and cold spots. These spots are the imprint of quantum fluctuations stretched across the sky. The largest structures in the cosmos trace their origins to the smallest possible quantum events. In this way, the uncertainty principle that governs individual particles also shaped the entire universe.

Although the quantum world is strange, it is not random or unpredictable in the way we might fear. It follows rules that can be described with great exactness using mathematical equations that have proven remarkably accurate. Those rules tell us how particles behave, how they interact, and how they combine to form everything we see. Scientists can calculate the properties of atoms, predict the outcomes of particle collisions, and design technologies that exploit quantum effects, all with extraordinary precision. The surprising behaviors of the quantum world are not accidents or glitches in reality. They are essential for the universe to exist in its current form. Without superposition, electrons couldn't occupy the multiple energy states that make chemistry possible. Without entanglement, certain reactions in photosynthesis and other biological processes might not work as efficiently. Without uncertainty, atoms would not form stable patterns, and the delicate balance that allows matter to exist would collapse. Without stable atoms, stars couldn't shine through nuclear fusion, planets couldn't form, and life couldn't appear. The quantum world, for all its strangeness, is the necessary foundation for everything familiar.

As we continue our journey through the history of the universe, we will see how these quantum rules allowed the first particles to form in the extreme conditions of the early universe and settle into stable structures as space cooled and expanded. The microscopic world, governed by laws we are only beginning to fully understand, created the foundation for the macroscopic world we live in. Quantum behavior is the invisible heartbeat of the universe, guiding the formation of matter and shaping the reality we experience. It operates beneath the surface of everything we touch and see, connecting the smallest particles to the largest cosmic structures. Understanding quantum behavior is not just an exercise in abstract physics. It is understanding the true nature of reality itself.

Chapter Six
Elementary Particles

The earliest moments of the universe created a landscape that was unfamiliar and extreme. Space was expanding rapidly, time had only just begun to flow, and the energy filling the universe was so intense that nothing resembling matter could exist in a stable form. Yet within this environment, the first ingredients of everything were beginning to take shape. These ingredients are what scientists call elementary particles. They are the smallest components the universe can build, and they cannot be broken down into anything simpler. Everything we see, from the light of distant stars to the cells in our bodies, is made from combinations of these tiny building blocks.

In the first fraction of a second after the universe began, the temperature exceeded ten trillion degrees. At such temperatures, the boundary between energy and matter dissolved completely. Particles and antiparticles appeared and vanished in a constant cycle, materializing spontaneously from the seething energy that filled space. These pairs were created from pure energy, lived for a brief moment, and then destroyed each other in bursts of radiation. This cycle happened millions of times in the blink of an eye. Each particle had a mirror twin, an antiparticle with opposite properties. When they met, they annihilated each other and returned to the energy from which they came.

As the universe expanded and cooled, this furious dance began to slow. The energy available for creating new particle pairs diminished with each passing moment. Some particles survived longer than others, and eventually, a small excess of matter over antimatter emerged. This asymmetry, slight but crucial, meant that when most particles and antiparticles destroyed each other, a tiny fraction of matter remained behind. The most basic of these survivors would become the foundation of all matter. These early survivors included quarks, electrons, and neutrinos. They floated in the growing space, waiting for conditions to change enough for them to join into more stable structures.

Quarks are among the most important particles the universe created. They come in six varieties, which physicists call flavors, but in the early universe only the lighter types were stable enough to persist. The two lightest quarks, known as up quarks and down quarks, became the primary constituents of ordinary matter. Quarks combine in specific ways to form protons and neutrons. A proton is made of two up quarks and one down quark, bound tightly together. A neutron is made of one up quark and two down quarks, arranged with slightly more mass. These combinations are held together by the strong force, which is one of the fundamental forces of nature and the most powerful force known in the universe.

The strong force works through particles called gluons, which constantly exchange between quarks and bind them with extraordinary strength. This binding becomes stronger as quarks try to separate, much like a spring that grows more resistant as it stretches. Because of this property, quarks can never be isolated or observed alone. They must always exist in combinations. Without this force, quarks would never join, protons and neutrons would never form, and atoms would be impossible. The birth of quarks and their ability to bind together was the first major step toward the creation of matter.

Electrons are another essential particle created in the early universe. They belong to a family of particles called leptons, which don't feel the strong force that binds quarks. Electrons have a very small mass compared to protons and neutrons, about two thousand times lighter, and they carry a negative electric charge. This charge is equal in strength but opposite in sign to the positive charge carried by protons. Electrons would eventually orbit around atomic nuclei, creating the structure of atoms and giving matter its chemical properties. In the early universe, electrons moved freely through the dense sea of radiation and matter, buffeted by photons and other particles. They were waiting for the universe to cool enough to allow them to attach themselves to the nuclei that would form later.

Neutrinos appeared during the same period. These particles are incredibly light, perhaps even massless by some measurements, and interact with matter so weakly that trillions of them pass through your body every second without leaving a trace. They have no electric charge and feel neither the strong force nor the electromagnetic force. They respond only to the weak nuclear force, which governs certain types of radioactive decay, and to gravity. In the early universe, neutrinos carried away energy and helped shape the behavior of matter during the first crucial seconds of expansion. Even though they rarely interact, their presence influenced the balance of particles that survived the heat of the beginning. They moved through the early universe almost like ghosts, shaping events without being easily detected.

Not all particles would go on to form matter. Photons, the particles of light, filled the early universe with a glow that was far too intense for atoms to exist. These massless particles of electromagnetic radiation carried energy in discrete packets, and they outnumbered the matter particles by billions to one. Photons constantly knocked electrons away from the particles they tried to join, scattering them before any stable bonds could form. The universe had to continue cooling until the energy dropped low enough for atoms to take shape. Until that moment came, the universe remained a hot, glowing plasma where matter and radiation mixed together in a dense, uniform environment. This plasma scattered light in every direction, making the universe opaque, much like the way water droplets in a thick fog scatter light and prevent us from seeing clearly.

The formation of elementary particles set the stage for everything that followed. These particles didn't choose their roles. Their properties, such as mass, charge, and the way they interact with forces, were fixed by the laws of physics. These properties determined how they behaved and what structures they could build. Protons and neutrons would join to form atomic nuclei. Electrons would orbit those nuclei to create atoms. Photons would later stretch into the faint cosmic glow that still fills the universe. Neutrinos would drift almost untouched through the cosmos. Every behavior, every structure, every piece of matter can be traced back to these particles and their interactions.

The properties of these particles are remarkable in their precision. If electrons were much heavier, atoms would collapse. If the strong force were slightly weaker, protons and neutrons couldn't hold together. If quarks had different masses, the universe might have produced only neutrons or only protons, and neither outcome would allow the chemistry necessary for complex structures. The universe needed this particular set of particles with this particular set of properties to build the world we observe.

As the universe continued to expand, its temperature fell enough for quarks to settle into protons and neutrons. When the universe was about one microsecond old, the temperature dropped below two trillion degrees. At this threshold, quarks could no longer roam freely. They locked themselves permanently into protons and neutrons through the strong force. These particles were finally stable. The universe had moved past the violent stage where everything appeared and disappeared too quickly to form anything lasting. The building blocks of atoms now existed, and the next chapter in the story would begin as these particles combined to create the first elements. The birth of elementary particles marks the moment when the universe gained the ingredients it needed to build everything from stars and planets to life itself.

Chapter Seven
How Atoms Formed

The birth of atoms was one of the most important turning points in the history of the universe. Before atoms existed, the universe was filled with a glowing mixture of particles and light that moved in wild, constant motion. Protons, neutrons, electrons, and photons drifted through space in a restless sea of energy, constantly colliding and breaking apart with tremendous force. Nothing was stable. Nothing could hold together long enough to form structure. The conditions were simply too extreme. The temperature was billions of degrees, far hotter than the core of any star that would ever exist. In this state, matter existed only as separate, unbound pieces. The universe had ingredients, but it had no recipe yet. To build stars, planets, and eventually life, the universe needed atoms. It needed particles that could stay together instead of falling apart. For that to happen, the universe had to cool, and it had to give matter the chance to settle into something calmer than the chaos of its beginnings.

A few minutes after the universe began, temperatures had dropped enough for protons and neutrons to start forming the first atomic nuclei. This process, called nucleosynthesis, happened remarkably quickly in cosmic terms. The universe was still unimaginably hot, around a billion degrees, but that was cool enough for the strong nuclear force to bind protons and neutrons together. These nuclei were simple, made mostly of hydrogen and helium. Hydrogen formed when a proton existed on its own with no neutron attached. Helium formed when protons and neutrons came together in small clusters, typically two protons bound with two neutrons. These clusters were the beginnings of the elements, but they were not yet atoms. Atoms require electrons orbiting their nuclei, and electrons were still moving freely through the universe, constantly knocked loose by the intense light that filled space. The nuclei had formed their cores, but they remained naked and incomplete, waiting for conditions that would allow them to capture electrons.

For hundreds of thousands of years, the universe remained too hot for electrons to join nuclei. Every time an electron approached a proton, drawn by the electromagnetic attraction between opposite charges, a photon of light would strike with enough energy to tear them apart before they could bond. The problem was not a lack of opportunity but a surplus of energy. The space was filled with such a bright and dense glow that matter couldn't settle into stable configurations. Light and matter were locked in a constant battle, with photons scattering off electrons in an endless series of collisions. This made the universe opaque, like a thick fog where light cannot travel far before bouncing off something. The universe was like a room so hot that ice could never form. Everything stayed in a loose, energetic state, waiting for the moment when conditions would finally allow stability. This period is sometimes called the plasma era, because matter existed in an electrically charged, ionized form rather than as neutral atoms.

As the universe continued to expand, its temperature dropped slowly and steadily. This cooling was a direct consequence of expansion itself. When space grows larger, the energy within it spreads out over a greater volume, and temperature falls as a result. The early fiery glow dimmed. Photons lost energy as their wavelengths stretched with the expanding fabric of space. Collisions became less violent. The universe was gradually losing the extreme conditions that had kept matter in its chaotic, unbound state. At around three hundred eighty thousand years after the beginning, the universe finally reached a temperature low enough for electrons to attach themselves to nuclei without being torn away. The temperature had dropped to about three thousand degrees, roughly half the temperature of the surface of the Sun. At this threshold, electrons could finally settle into stable orbits around nuclei. This event is known to scientists as recombination, though the term is somewhat misleading since electrons and nuclei were combining for the first time rather than recombining. It was the moment when the universe became transparent. Light could travel freely for the first time because it was no longer constantly scattered by loose electrons. The photons that had been trapped in endless collisions were suddenly released to move through space unimpeded. The glow of that first free light still exists today as the cosmic microwave background radiation, a faint echo of the moment when atoms appeared. This ancient light, now cooled to just a few degrees above absolute zero, fills the universe and provides a snapshot of what the cosmos looked like at that crucial transition.

When electrons attached themselves to the waiting nuclei, the first true atoms formed. Hydrogen atoms were the simplest and most common. A single proton with a single electron became the most abundant element in the universe, making up about three quarters of all ordinary matter by mass. Helium atoms, made of two protons and two neutrons with two electrons, were the next most common, accounting for most of the remaining quarter. A small amount of lithium and other light elements also appeared during those first few minutes of nucleosynthesis, but almost everything was hydrogen and helium. These proportions were set by the conditions of the early universe and have remained roughly constant ever since. These atoms filled the universe, drifting silently in the growing emptiness of space, now transparent and calm compared to the brilliant fog that had preceded them.

Atoms are remarkable because they are stable yet flexible. The nucleus, tiny and dense, holds most of the mass in an incredibly small space at the center. Meanwhile, electrons occupy regions around the nucleus that are vastly larger by comparison, forming a cloud of probability where the electron is likely to be found. These electron clouds determine how atoms interact with one another. When atoms come close together, their electron clouds can overlap, share, or repel each other, creating the bonds and forces that give rise to chemistry. Chemistry, in turn, makes complex structures possible. The formation of atoms was the beginning of matter as we know it. Before atoms, there could be no molecules, no chemical reactions, no way for matter to organize itself into anything more intricate than loose particles. Atoms became the seeds that gravity would later pull together into clouds, then into stars, and then into galaxies. Without atoms, the universe would have remained a smooth sea of particles and energy with no variation, no structure, and no possibility for the emergence of complexity or life.

In the calm that followed the formation of atoms, the universe entered a new phase. The intense glow faded into darkness. For the first time, the universe was truly dark, with no stars yet burning to provide light. Vast regions of hydrogen and helium gathered slowly under the influence of gravity. Though gravity is the weakest of the fundamental forces, it acts over infinite distances and never cancels out, making it the dominant force on cosmic scales. These clouds of neutral atoms began to feel the pull of their own mass. They grew denser over time as gravity drew more and more matter into regions where the density was already slightly higher. Where the density became high enough, the next major chapter of cosmic evolution began. Atoms were no longer drifting alone in isolated independence. They were beginning to gather, to collapse inward, and to create the conditions necessary for the first stars. The universe was preparing to light itself for the first time, ending what astronomers call the cosmic dark ages.

The appearance of atoms marks the end of the universe's earliest era and the beginning of everything familiar. From this moment on, matter could build the structures that would fill the cosmos. The transition from an opaque plasma to a transparent gas of neutral atoms was more than just a change in temperature. It was a fundamental shift in the nature of the universe, opening the door to complexity and evolution. The simple atoms of hydrogen and helium that formed in this era still make up most of the visible matter in the universe today. They are the silent witnesses to the long journey from pure energy to the complex world we now inhabit. Their formation was the moment when the universe gained the stability and structure it needed to move toward the creation of stars, planets, and eventually life. Everything that came after depended on this seemingly simple achievement: the binding of electrons to nuclei, creating the first stable atoms in the cooling darkness of an expanding cosmos.

Chapter Eight
How Stars Made the Elements

When atoms first formed, the universe became calmer, but it was still dark. There were no stars, no planets, and no sources of light except the fading glow left over from the beginning. Hydrogen and helium drifted through space in enormous clouds that stretched across vast distances. These clouds were almost perfectly smooth, but not completely. Tiny variations in density, created in the earliest moments of the universe, left some regions slightly thicker than others. Over time, gravity pulled on those regions, causing gas to collect. The denser areas exerted a stronger gravitational pull, drawing in more material from their surroundings. This created a feedback loop where dense regions grew denser still. Little by little, the first structures began to appear as these clouds gathered into denser knots. These knots would become the first stars.

As these early clouds collapsed, gravity squeezed the atoms together. The pressure increased, and the temperature rose. The gravitational energy of the infalling gas converted into heat, warming the center of each cloud. Deep in the center of each growing cloud, heat built until the atoms could no longer behave the way they did in ordinary space. When the temperature reached millions of degrees, something remarkable happened. The thermal energy became so intense that when hydrogen nuclei collided, they moved fast enough to overcome the electromagnetic force that normally pushes positively charged particles apart. Hydrogen atoms fused together. This fusion released energy in the form of light and heat, and that energy pushed outward against the surrounding gas, balancing the inward pull of gravity. The star ignited. For the first time in the universe, light was born from matter itself. Countless stars formed across the cosmos over millions of years, filling the darkness with their glow. The universe had entered a new era, one illuminated by the fires of fusion.

Inside a star, atoms behave differently than they do here on Earth. The heat and pressure at the center are so intense that electrons cannot stay bound to their nuclei. The gas becomes a hot, charged mixture that scientists call plasma. In this state, atomic nuclei stripped of their electrons move freely, colliding with tremendous speed. In this environment, nuclei collide with enough force to overcome their natural tendency to repel each other. The positive charges that would normally push nuclei apart are defeated by the sheer kinetic energy of the particles. When they collide and join, they form a new, heavier nucleus, and in the process they release energy according to Einstein's famous equation relating mass to energy. A tiny amount of mass converts into a tremendous amount of energy. This process is called nuclear fusion. Fusion is what makes stars shine, and it is the process that created most of the elements found in the universe.

The simplest fusion reaction happens when hydrogen nuclei combine to form helium. In the cores of stars, hydrogen nuclei collide and merge through a sequence of steps, eventually creating helium nuclei while releasing energy in the form of photons and fast-moving particles. This was the first step, and it happened throughout the early stars. Over time, as stars used up their hydrogen, their cores contracted under gravity. This contraction raised the temperature and pressure once more until conditions became extreme enough to fuse helium into heavier elements. Helium nuclei could collide to form carbon through a delicate process that requires precise conditions. Carbon could become oxygen when it captured additional helium nuclei. Oxygen could become neon and magnesium through further collisions. These steps continued inside many stars, slowly building up the elements that would become essential for planets and life. Each fusion reaction released energy that kept the star shining and maintained the balance between gravity's inward pull and the outward push of radiation pressure. The star became a kind of forge, transforming light elements into heavier ones through the intense heat of its core.

Not all stars live the same kind of life. Small stars burn gently and can last for billions of years, slowly turning hydrogen into helium. Their lower temperatures mean they work through their fuel at a measured pace. They may never reach the conditions needed to forge elements much heavier than helium. Large stars burn hotter and faster. Their greater mass means stronger gravity, which compresses their cores to higher temperatures and pressures. They build heavier elements in their cores, adding layers like the structure of an onion. Each layer burns a different element at a different temperature. Hydrogen burns in the outermost shell. Below that, helium fuses into carbon and oxygen. Deeper still, carbon burns to make neon and magnesium. Oxygen fuses to create silicon and sulfur. These layers contain different elements forming at different temperatures, with the coolest reactions happening in the outer shells and the hottest at the very center. In the center, the heaviest elements appear as the star approaches the end of its life. Silicon, the final fuel, fuses to create iron. But there is a limit. Iron occupies a special place on the periodic table. Its nucleus is so tightly bound that fusing it into heavier elements requires more energy than it releases. When a star tries to fuse iron, the process no longer releases energy. The nuclear reactions that have supported the star for millions of years suddenly stop producing the outward pressure needed to resist gravity. Instead of shining brighter, the star loses its support. Gravity wins.

When a massive star can no longer produce energy through fusion, its core collapses inward with incredible force. In less than a second, the iron core implodes, compressing matter to densities found nowhere else in the universe except in neutron stars and black holes. The outer layers fall with it, then rebound outward in a powerful explosion when they strike the incompressibly dense core. This explosion is called a supernova. It is one of the most violent events in the universe, bright enough to outshine an entire galaxy for a brief moment. In that moment, the star releases as much energy as it produced during its entire lifetime. During a supernova, temperatures soar to billions of degrees, high enough to forge the heaviest elements in existence. The flood of neutrons created in the explosion bombards lighter nuclei, building them up through rapid capture processes that add mass far faster than ordinary fusion ever could. Gold, uranium, platinum, and many other rare elements were created in these final seconds of a dying star. They were scattered across space in enormous clouds of debris traveling at thousands of kilometers per second. The explosion enriched interstellar space with the products of stellar alchemy. Everything heavier than iron owes its existence to these explosions or to similar processes in colliding neutron stars.

The material thrown out by supernovas became part of the next generation of stars and planets. New clouds of gas and dust formed from the remains of older stars, now enriched with the full periodic table of elements. Gravity collected these materials into new star systems. When these enriched clouds collapsed, they formed not only stars but also the planets, moons, and asteroids that orbit them. The planets, moons, and asteroids that formed from these clouds contained the full range of elements created inside stars. Unlike the first generation of stars, which formed from nearly pure hydrogen and helium, these later generations had access to carbon, oxygen, silicon, iron, and all the other elements needed to build rocky worlds. The carbon in your body, the calcium in your bones, the oxygen you breathe, and the iron in your blood all began inside ancient stars that lived and died long before the Sun was born. You are made of stardust in the most literal sense.

Stars are often described as factories because they transform simple atoms into complex elements. This description captures their essential role in cosmic evolution. Without stars, the universe would still be filled with only hydrogen and helium. There would be no planets, no rocks, no oceans, no air, and no life. Stars created the ingredients needed for everything else. Their formation, life, and death shaped the chemical richness of the universe. The process continues today. Even now, stars continue to build new elements in their cores, and supernovas continue to spread them across space. Each generation of stars enriches the cosmos a little more, adding to the diversity of matter available for future generations.

The story of the elements is the story of how the universe learned to build complexity. Simple atoms came first, forged in the hot early moments after the beginning. But stars gave the universe a wider palette of materials, transforming those simple ingredients into the diverse elements we see today. With these materials spread across space, carried on the shockwaves of exploded stars, the next chapter could begin. Clouds of enriched gas would gather once more under gravity's patient pull, this time forming planetary systems with the chemical variety needed for richer possibilities. In these systems, chemistry would grow more intricate, and molecules would begin to appear. The universe was preparing not only to build worlds, but eventually to build life.

Chapter Nine
How Molecules Formed

The universe became a richer place once stars had created the elements, but those elements were still scattered as single atoms drifting through space. Atoms alone cannot create the complexity needed for planets, water, air, or life. A lone oxygen atom cannot hold liquid. A single carbon atom cannot form a structure. To build anything with substance and pattern, atoms must join together through chemical bonds, sharing or exchanging electrons to create stable partnerships. When atoms join, they form molecules, and molecules are the first step toward chemistry. The appearance of molecules marks the moment when the universe learned to combine its ingredients in new ways, creating patterns and shapes that would eventually lead to living systems.

The earliest molecules formed long before planets existed. They appeared in the vast clouds of gas that filled young galaxies, regions where temperatures had dropped to a few hundred degrees above absolute zero. These clouds were colder and calmer than the centers of stars, and in these gentler conditions atoms could interact without being torn apart by intense radiation or heat. Hydrogen was the most abundant element, and it formed the first molecule in the universe when two hydrogen atoms joined. This molecular hydrogen, known as H2, held together through a simple covalent bond where two electrons orbit both nuclei. The formation of this bond released a small amount of energy, and this energy could escape the gas cloud as infrared radiation. This simple molecule helped cool gas clouds by allowing them to lose energy more efficiently than individual atoms could. As the clouds cooled, they contracted under their own gravity more easily, compressing until conditions were right for new generations of stars to ignite. Even the simplest molecules played a role in shaping the structure of galaxies, acting as thermostats that regulated the birth of stars.

As new stars formed, they flooded their surroundings with ultraviolet light strong enough to break many molecules apart, splitting their bonds and returning them to individual atoms. Yet the outer regions of galaxies remained cool and quiet, shielded from the most intense starlight. In these protected zones, atoms and molecules collected on tiny grains of dust floating in space. These dust grains were themselves remarkable structures, often composed of carbon or silicates formed in the atmospheres of dying stars. They measured less than a micron across, but they acted like meeting places, providing surfaces where atoms could stick together long enough to react. On Earth, molecules form when atoms collide with enough energy to overcome repulsion between their electrons. In the near vacuum of space, such collisions are rare. Dust grains solved this problem by holding atoms in place, allowing them time to bond before drifting apart.

Oxygen combined with hydrogen on these grain surfaces to form water, molecule by molecule. Carbon joined with hydrogen to form methane and other simple hydrocarbons. Carbon monoxide appeared when carbon and oxygen met. Nitrogen bonded with hydrogen to create ammonia. Sulfur, phosphorus, and other elements participated in this early chemistry. Each reaction added to the inventory of molecules floating through space. Space was becoming a chemical workshop, filled with building blocks that would one day be important for life. These molecules accumulated in dense clouds spanning light years, regions so thick with gas and dust that visible light couldn't penetrate. Within these dark clouds, shielded from destructive radiation, molecules could survive and grow more complex.

This early molecular activity didn't create life on its own. It created diversity. A wide range of molecules formed in small protected regions of space, and these molecules accumulated in clouds that would eventually collapse into new solar systems. Astronomers have now detected over two hundred different molecules in space, ranging from simple diatomic pairs to complex organic molecules containing ten or more atoms. Some contain long chains of carbon. Others feature rings or branches. This molecular diversity meant that when clouds collapsed under gravity, they brought a varied chemical palette with them. When these clouds fell inward, they formed young stars surrounded by flattened disks of gas and dust, swirling pancakes of matter that would become solar systems. Inside these protoplanetary disks, temperatures varied dramatically. Near the infant star, heat kept everything vaporized. Further out, temperatures dropped low enough for water to condense into ice. At the outermost edges, even gases like methane and ammonia froze solid.

These temperature differences allowed even more complex molecules to form through a process chemists call differentiation. Close to the star, only refractory molecules that could withstand heat survived. In the cooler middle regions, water ice accumulated along with carbon dioxide and other volatile compounds. In the coldest outer zones, a rich mixture of organic molecules collected, protected by frozen mantles of ice. Some of these organic molecules contained chains of carbon atoms bonded in repeating patterns. Others featured functional groups, clusters of atoms that gave molecules specific properties and allowed them to participate in particular reactions. They were stable enough to survive the formation of planets and the violent collisions that assembled worlds from smaller pieces.

When young planets finally formed from this swirling disk material, they inherited these molecules. Some were locked inside icy comets and asteroids, frozen libraries of early chemistry preserved for billions of years. Others arrived through meteor impacts, delivered to planetary surfaces in spectacular collisions that also brought water and minerals.

Still others formed on the planets themselves once conditions allowed. On worlds with atmospheres, lightning could provide the energy to break simple molecules apart and reassemble them into new combinations. On worlds with liquid water, chemistry became more active and versatile. Water acted as a solvent, helping molecules collide and interact more easily than they could in the dry vacuum of space or locked in ice. Dissolved in water, molecules could move freely, bumping into one another millions of times per second. Heat from volcanic activity provided energy for reactions. Ultraviolet light from the sun broke bonds and created reactive fragments that could recombine in novel ways. Earth, like other rocky planets in the habitable zones of their stars, became a place where simple molecules could combine into more complex structures.

Molecules are remarkable because they allow matter to build patterns that atoms alone cannot create. A single atom is limited in what it can do. It can absorb or emit light at specific wavelengths. It can lose or gain electrons to become an ion. But a molecule containing several atoms can fold into precise three-dimensional shapes. It can twist and flex. It can store energy in its bonds and release that energy when bonds break. The shape of a molecule affects how it behaves, determining which other molecules it can interact with, much like how the shape of a key determines which lock it can open. As molecules grew more complex, they gained the ability to take part in more elaborate reactions. Long chains of carbon formed, creating molecules with dozens or hundreds of atoms linked in sequence. Rings of carbon created stable structures that could serve as frameworks for attaching other groups of atoms. Molecules began to connect in repeating patterns, building lattices and networks that hinted at the possibility of self-organization. The universe was not yet alive, but it was building the tools that life would later use.

The formation of molecules is a quiet part of the universe's story, but it is one of the most important. Molecules gave matter flexibility. They allowed reactions to occur at temperatures far below the millions of degrees needed for nuclear fusion. They allowed energy to be stored in chemical bonds and transferred from one molecule to another. They created variety, turning a handful of elements into millions of different substances. They formed the foundation for biology. Without molecules, there would be no water to fill oceans, no oxygen to breathe, no carbon dioxide to sustain plant life, no soil to support roots, and no living creatures to inhabit any world. Molecules are the bridge between simple matter and complex systems. They make chemistry possible, and chemistry makes life possible.

As the universe continued its long path toward complexity, molecules gathered in oceans, clouds, and rocky environments. They reacted, broke apart, and reformed countless times, driven by heat, light, and the constant motion of matter. The molecular inventory of planets grew richer with time. Eventually, through processes that required both chance and the laws of chemistry, some molecules gained the ability to copy themselves, creating identical versions through chemical reactions. Others developed the capacity to speed up reactions, acting as catalysts that helped guide chemical processes along specific pathways. These developments, emerging from the random exploration of chemical possibilities, set the stage for one of the greatest transitions in the history of the universe. The next chapter of the story begins when molecules learned to organize themselves into the first living cells.

Chapter Ten
The Chemistry of Life

Life didn't appear suddenly or by chance. It emerged from a long chain of chemical steps that began with simple molecules in a young and restless world. When Earth first formed around four and a half billion years ago, it was a place of extremes. Volcanoes released thick clouds of gas into a sky that held no oxygen. Lightning flashed through a heavy atmosphere rich in carbon dioxide, nitrogen, methane, and water vapor. Meteorites delivered new materials from space, bringing with them organic compounds formed in the cold darkness between stars. The oceans simmered with heat from below, warmed by radioactive decay in the young planet's core and by hydrothermal vents that pumped mineral-rich water into the depths. Sunlight poured down from above, carrying ultraviolet radiation that could both break molecules apart and drive new reactions. In this harsh but chemically active environment, molecules already present on Earth began to interact in ways that were more complex than anything that had happened before. The chemistry of life began slowly, but with steady progress, guided by the laws of nature.

The early Earth contained water, carbon, nitrogen, sulfur, phosphorus, and many other elements created long ago inside stars. These elements were not rare or unusual. Carbon, the foundation of all organic chemistry, forms bonds easily and in many different arrangements. Hydrogen and oxygen, the components of water, were abundant. These elements combined into simple organic molecules through various pathways. Some arrived on meteorites and comets, delivered intact from elsewhere in the solar system. Others formed in the atmosphere when lightning or ultraviolet light provided the energy needed to break apart simple gases and recombine them into larger molecules like amino acids, the building blocks of proteins, and nucleotide bases, the components of genetic material. Many of these newly formed molecules settled into the oceans where currents stirred them constantly, bringing reactants together in an endless mixing process. In these waters, molecules drifted, collided, and reacted. Some reactions were driven by heat from hydrothermal vents on the sea floor, where temperatures could exceed the boiling point of water under pressure. Others were powered by sunlight in shallow pools where evaporation concentrated dissolved chemicals. Over millions of years, these reactions produced a wide variety of molecules, and some of these molecules had properties that helped them survive and accumulate rather than breaking down immediately.

Among these molecules were those capable of holding information in a chemical form. They were long chains built from repeating units, much like a string of beads where each bead could be one of several types. Molecules with specific sequences of units formed sections that could attract complementary partners through weak chemical bonds. When one chain met another with a matching pattern, they stuck together temporarily. These patterns were not yet instructions in the way DNA works today, but they were early forms of chemical memory, a way for molecules to recognize and interact with specific partners rather than reacting randomly with anything nearby. Other molecules acted as catalysts. They sped up reactions without being consumed themselves, making certain interactions thousands or even millions of times more likely than they would be otherwise. Some of these catalysts were small metal-containing molecules. Others were chains that folded into particular shapes, creating pockets where specific reactions could occur. Together, information molecules and catalytic molecules created a small but important advantage in the primordial soup. They allowed certain chemical pathways to repeat themselves in cycles rather than ending after a single reaction. A catalyst could help an information molecule copy itself. That copy could then help create more catalyst.

This created a feedback loop, a self-reinforcing system where success bred more success.

Life requires organization, and organization requires a way for molecules to stay close together rather than diffusing into the vast ocean. On the early Earth, this became possible when fatty molecules formed simple membranes. Fatty molecules have two distinct ends. One end attracts water while the other repels it. When these molecules are placed in water, they spontaneously arrange themselves into double layers with their water-repelling ends hidden inside and their water-attracting ends exposed to the surrounding liquid. These membranes were not advanced or carefully designed, but they could trap water and dissolve useful molecules inside while keeping out others. When a membrane closed on itself, perhaps stirred by currents or waves, it created a small bubble called a vesicle. Inside that bubble, the molecules of early chemistry were held together in one place instead of drifting apart into dilution. These bubbles acted like the earliest containers of life, primitive versions of the cell membrane that surrounds every living thing today. They kept helpful molecules close enough to interact frequently and allowed them to evolve together as a system. A reaction that produced useful products inside a vesicle benefited only that vesicle, not the entire ocean. This created the conditions for chemical evolution to begin in earnest.

Slowly, over countless generations of molecular interactions, the chemistry inside these bubbles grew more complex and more coordinated. Information molecules began to copy themselves more reliably as catalysts became better at assisting the copying process and correcting errors. Catalytic molecules became more efficient at guiding reactions, folding into shapes that were perfectly suited to speed up specific transformations. Membranes became more stable and selective, allowing useful molecules to enter while keeping harmful ones out, and even developing primitive ways to capture energy from the environment. These improvements didn't happen because the molecules had goals or intentions. They happened because the combinations that stayed together and functioned effectively were more likely to continue reacting and copying than those that fell apart or worked poorly. When a vesicle contained a particularly good combination of molecules, it could grow by absorbing more fatty molecules from the surroundings. Eventually it might split into two vesicles, each containing roughly the same molecular machinery. Over immense stretches of time, the successful combinations became more common while less effective ones disappeared. Chemistry was moving toward biology through a process of natural selection that operated at the molecular level.

The first true cells were simple by modern standards, but they represented a profound transition. They contained strands of genetic material that could store information and pass it to offspring. They held a collection of catalytic molecules that worked together to extract energy from the environment and use it to build new molecules. They possessed a membrane that could grow, divide, and regulate what entered and left the cell. These early cells were not yet alive in the way we think of plants or animals with their elaborate structures and behaviors, but they were self-sustaining systems capable of the three defining features of life. They could take in energy from their surroundings and use it to maintain their internal organization against the natural tendency toward disorder. They could grow by building copies of their molecular components. They could reproduce by making copies of themselves, passing their genetic information to the next generation. When a cell divided, two new cells appeared, each containing the same molecules and the same information as the original. With each generation, small changes appeared due to errors in copying or variations in conditions, and some changes helped the cells survive better than others. A cell that could extract energy more efficiently would grow faster. A cell with more stable genetic material would pass on its traits more reliably. These changes, combined with the ability to copy and reproduce, created the first steps of biological evolution. The chemistry of life had given rise to living things.

The chemistry of life is remarkable because it turns simple interactions between mindless molecules into systems that can grow, adapt, and respond to their environment. Carbon forms the backbone of complex molecules because it can bond to four other atoms at once, creating chains, branches, and rings of virtually unlimited variety. Water allows reactions to take place easily because it dissolves so many different substances and brings them together in solution, and because water molecules themselves can participate in reactions by donating or accepting hydrogen atoms. Nitrogen and phosphorus help build the structure of genetic material, with nitrogen forming the bases that spell out genetic information and phosphorus linking those bases into long stable chains. Metals found in rocks and dissolved in seawater become part of enzymes that speed up reactions, with iron, magnesium, zinc, and other elements sitting at the active centers of these molecular machines. Light provides energy for some early life forms through simple photochemical reactions where light-absorbing molecules capture photons and use their energy to drive the synthesis of useful compounds. Heat from the Earth, whether from volcanic activity or hydrothermal vents, provides energy for others by creating temperature gradients that can be harnessed to move molecules and power reactions. Every part of chemistry, every element and every type of reaction, contributes to the process of life.

As life continued to evolve over hundreds of millions of years, chemical systems grew more elaborate and more refined. Molecules learned, through the trial and error of natural selection, to store information with greater accuracy using the stable double helix structure we see in DNA today. Cells developed sophisticated ways to use energy more efficiently, creating molecular machines that could capture energy from sunlight or chemical reactions and store it in forms that could be used later. Membranes became stronger and more complex, embedded with proteins that could transport specific molecules in and out, sense conditions in the environment, and communicate with other cells. Within these changes, life found new possibilities and explored new strategies for survival. Over vast stretches of time measured in billions of years, the simple chemistry of early Earth produced the first living organisms. From those organisms came the incredible variety of life that fills the planet today, from bacteria that thrive in boiling water to plants that harvest sunlight to animals that think and dream.

The chemistry of life shows us something important about the universe and our place in it. Life is not separate from nature or exempt from physical laws. It is a natural outcome of the laws that govern matter and energy, an expected result when the right conditions are met. When the right molecules form in the right environment and are given enough time to interact and evolve, they can create systems that copy themselves, change over generations, and grow in complexity. Life is the universe exploring what it can build from itself, starting with the smallest molecules and moving toward the complexity of living beings with awareness and purpose. With the appearance of the first cells, the story of biology began in earnest, and the universe stepped into a new chapter where life would learn to evolve, adapt, and eventually understand its own origins. The chemistry that began in a violent young world led to something extraordinary, proving that the potential for life was written into the fabric of the cosmos from the very beginning.

Chapter Eleven
The First Cells

Once the chemistry of life found a stable footing, the universe crossed an invisible boundary. Molecules that had passed energy, copied fragments of themselves, and reacted in repeating cycles finally organized into the first true cells. These early cells were simple, but they represented a major shift in the story of the universe. For the first time, matter could repair itself, reproduce, and evolve in a way that allowed changes to accumulate over generations. Before cells, chemical reactions happened and then ended. The products drifted apart, and any useful pattern was lost. With cells, patterns could be preserved, refined, and passed forward. The rise of the first cells marked the moment when the universe learned how to create living systems that could persist across time and respond to their surroundings.

The earliest cells were small structures surrounded by thin membranes. These membranes formed naturally when certain molecules found themselves in water. Some molecules have one end that binds easily to water and another end that pushes water away. In the right conditions, these molecules arrange themselves into double layers, creating a barrier that separates inside from outside. This simple boundary became one of the most important inventions in the history of life. The membrane kept useful molecules inside while letting energy and simple nutrients enter from the environment. It created an interior space where conditions could be controlled and where reactions could happen in sequences rather than at random. Inside each cell, chains of genetic material guided basic reactions. These chains held patterns that could be copied, although the copying was not yet very accurate. The molecules that did the copying had to read the pattern and assemble a matching version, step by step. Mistakes were common, but these mistakes were essential. The copying machinery sometimes grabbed the wrong building block or skipped a section entirely. Some changes made no difference to how the cell functioned. Others damaged critical processes and prevented the cell from surviving. A few rare changes made the cell slightly better at surviving or copying, perhaps by allowing it to use a new source of food or resist a harmful chemical. Over time, these small advantages spread through populations as successful cells divided and unsuccessful ones faded away. This was natural selection at its simplest, working on the smallest units of life the universe had yet produced.

Early cells lived in a world that was far more dangerous and unpredictable than Earth is today. Volcanoes released hot gases into the air, filling the sky with carbon dioxide, nitrogen, and sulfur compounds. The atmosphere lacked oxygen and was filled with chemicals that would be harmful to modern life. Lightning storms crackled through thick clouds, and the surface was pelted by rocks from space that had not yet been swept up by the planets. Ultraviolet light from the Sun reached the surface in full strength because Earth had no protective ozone layer to filter it. This radiation was powerful enough to break apart the delicate molecules that living cells depended on. Yet the oceans provided shelter. Water absorbed harmful radiation and gave cells a stable environment where they could grow and divide. Even a few meters below the surface, the intensity of ultraviolet light dropped dramatically. Temperature changes that would have been extreme on land were smoothed out by the enormous heat capacity of liquid water. Dissolved minerals provided raw materials for building cellular structures. In the quiet of these oceans, protected from the chaos above, early life found its first home and began to spread.

These first cells were also the first organisms in the truest sense. They fed on chemicals in the water and used simple reactions to gain energy. Some cells used energy from the environment to build more complex molecules from simpler parts, constructing the structures they needed to grow. Others broke down molecules they found in their surroundings to release energy they could use for maintenance and reproduction. These different approaches reflected the variety of chemical environments available on early Earth. Near hydrothermal vents, where hot mineral rich water poured up from the ocean floor, cells could harvest energy from reactions between those minerals and the cold seawater. In shallower regions, dissolved chemicals delivered by rivers or released from rocks provided different opportunities. This metabolic flexibility gave early life an important advantage. If one source of energy became scarce or disappeared entirely, another might still be available. Cells that could switch between strategies, or that could exploit resources others ignored, were more likely to survive when conditions shifted. This flexibility helped life endure the unstable conditions of early Earth and gave evolution more paths to explore.

As cells continued to divide, they formed communities. Groups of cells lived together in thin layers across rocks, in deep ocean vents, and in shallow pools warmed by sunlight. These communities were not random gatherings. Cells that stayed close to one another began to interact in ways that affected their survival. These communities created small ecosystems where different types of cells supported one another, even if unintentionally. Some cells produced simple molecules as waste products of their metabolism. These waste molecules drifted into the surrounding water where other cells, with different chemical abilities, could use them as fuel or building materials. What was garbage to one cell became food for another. Cooperation and competition began to shape the direction of evolution in ways that went beyond the survival of individual cells. Cells that lived near helpful neighbors thrived. Cells that produced something others needed attracted partners. Cells that competed for the same scarce resource had to become more efficient or find something new. Life was slowly becoming more organized and more capable of adjusting to its environment. The individual cell was no longer the only unit that mattered. Groups and relationships began to matter too.

One of the most important changes in early life happened when certain cells learned how to use sunlight for energy. This development took time and required many evolutionary steps. These cells developed structures containing special molecules that could capture light and convert it into chemical energy. When light struck these molecules, it knocked electrons into higher energy states. The cell then guided those electrons through a series of steps, extracting energy at each stage and storing it in a usable form. This process eventually produced oxygen as a byproduct when cells began splitting water molecules to replace the electrons they had lost. At first, oxygen was toxic to many living organisms. It reacted aggressively with the molecules inside cells, damaging the delicate machinery of life. The oceans absorbed much of this early oxygen, pulling it out of the water by reacting with dissolved iron and other metals. For hundreds of millions of years, oxygen levels remained low. But over time, some cells evolved defenses against oxygen and then learned to use oxygen in their energy systems. This turned a deadly poison into a powerful tool. Oxygen allowed cells to extract far more energy from food molecules than earlier methods had permitted. A cell using oxygen could thrive on smaller amounts of food and grow more complex without running out of energy. The presence of oxygen in the atmosphere changed the planet forever. It made the environment more stable by creating the ozone layer that blocked ultraviolet radiation. It altered the chemistry of rocks and oceans. Most importantly, it created new opportunities for evolution by opening up energy intensive pathways that had been impossible before.

Even with these advances, early life remained simple for many millions of years. Cells divided, evolved, and adapted to a changing Earth, but they did so slowly by the standards of later evolution. There were no eyes to see with, no legs to walk on, no thoughts or intentions. There was only chemistry, running in loops that preserved themselves by making copies. Yet this simplicity was also a kind of success. These cells spread into every environment that could support them, from the coldest depths of the ocean to the warmest shallow pools. In some places, life formed mats that covered rocks and ocean floors, building up in layers as generations lived and died on top of one another. In others, it filled the water with microscopic activity, turning clear seas cloudy with living things. These cells were the ancestors of every living thing on the planet. Every plant, animal, and human can trace its family line back to these small, ancient organisms. The genetic code we carry, the way our cells produce energy, the basic chemistry of metabolism, all of these were established in those early times and have been passed down through billions of years of evolution. We are, in a very real sense, the distant descendants of those first simple cells.

The rise of early life shows how much the universe can accomplish with simple tools. A cell is made of common molecules found throughout the cosmos: carbon, hydrogen, oxygen, nitrogen, phosphorus, and a few others. None of these elements are rare. None of them have special properties that exist only on Earth. Yet when combined in the right way, guided by the laws of chemistry and physics, they become capable of growth, adaptation, and survival. They become able to respond to their environment, to change over time, and to produce the stunning diversity of life we see today. From these early beginnings, life gained the stability it needed to explore new forms. The membrane provided protection. Genetic material provided memory. Metabolism provided energy. Reproduction provided continuity. With these foundations secure, evolution could begin to experiment. The next steps would take life beyond single cells, leading to organisms that were larger, more complex, and able to move and sense their surroundings in ways the first cells never could.

With the foundation of the first cells in place, evolution had everything it needed to continue its long journey. The basic principles had been established. Variation, selection, and inheritance would continue to operate, generation after generation, shaping life in response to the challenges and opportunities each environment presented. Life would not remain small forever. In the chapters to come, we will see how simple cells gradually gave rise to more complex organisms, how cooperation between cells led to entirely new forms of life, and how those organisms eventually shaped the planet we call home.

Chapter Twelve
The Rise of Oxygen

Life on early Earth began in a world that would feel completely unrecognizable to us. The air contained almost no oxygen. Instead, the atmosphere held a dense mixture of gases that would be poisonous to most modern organisms. Carbon dioxide filled the air at concentrations hundreds of times higher than today. Methane drifted through the lower atmosphere, sometimes forming hazy layers that trapped heat near the surface. Nitrogen was present, much as it is now, but the balance of everything else was profoundly different. The oceans were rich with dissolved iron, giving them a greenish tint in some regions, and the sky may have carried a faint reddish or orange hue from volcanic gases and the scattering of light through an atmosphere thick with particles. The surface was shaped by constant geological activity, with volcanoes releasing carbon dioxide, methane, sulfur compounds, and water vapor in enormous quantities. Into this environment emerged the first simple cells, living quietly in a planet that didn't yet resemble the world we know.

These early organisms survived on chemical reactions that didn't require oxygen. They drew energy from compounds like hydrogen sulfide, iron, and various sulfur-bearing minerals dissolved in the water. Some lived near hydrothermal vents on the ocean floor, where hot, mineral-rich water provided the raw materials for their metabolism. Others floated in shallow waters, harvesting energy from simpler chemical gradients. For millions of years, oxygen was not a part of their environment. It was not a resource they could use. It was, in fact, a threat. Oxygen is a highly reactive element. It binds readily to other molecules, disrupting delicate structures and breaking apart the chemical machinery that early cells relied upon to survive. In a world without oxygen, life had evolved intricate systems that couldn't tolerate its presence.

The rise of oxygen didn't begin with fire or plants or anything familiar from the modern world. It began with single-celled organisms drifting in the oceans, likely in shallow coastal waters where sunlight could penetrate. These organisms were ancestors of modern cyanobacteria, and some of them developed a remarkable ability through gradual evolutionary changes. They learned how to use sunlight to power chemical reactions, splitting water molecules to extract hydrogen and leaving oxygen behind. This process, known today as photosynthesis, began slowly and quietly, perhaps as early as three billion years ago. The earliest photosynthetic organisms were not efficient by modern standards. They produced oxygen in small amounts, and the planet was not yet ready to hold it.

At first, the oxygen produced by these early bacteria reacted with the environment around them almost as quickly as it was made. It combined with dissolved iron in the oceans, forming iron oxide in a reaction similar to rusting. This iron oxide settled to the ocean floor in layers, building up over millions of years into distinctive banded iron formations. These formations, striped with alternating layers of iron-rich and iron-poor rock, still exist today and can be found on every continent. They serve as a geological record of a time when oxygen was being produced but couldn't yet accumulate freely. The oceans acted as a vast chemical buffer, absorbing oxygen immediately and preventing it from reaching the atmosphere. Other elements played similar roles. Sulfur compounds and volcanic gases reacted with oxygen, consuming it before it could escape into the air.

Over time, something changed. The amount of oxygen entering the oceans increased as photosynthetic bacteria spread and became more efficient. The supply of dissolved iron began to run out. There was only so much iron available in the water, and once it had all been oxidized and settled out, the oceans could no longer absorb all the oxygen produced by photosynthetic bacteria. Gradually, oxygen started to accumulate in the atmosphere. This shift was not sudden. It happened over hundreds of millions of years, punctuated by periods of stability and periods of rapid change. The effects, however, were profound and irreversible. The presence of oxygen reshaped the chemistry of the entire planet. Minerals formed that couldn't have existed before. Red beds of oxidized sediment began to appear on land, marking the transition to a world where oxygen was present at the surface. The atmosphere gained new properties, altering the behavior of weather and climate in ways that would have cascading effects. The rise of oxygen transformed Earth from a reducing environment, where chemicals tended to donate electrons, to an oxidizing one, where chemicals readily accepted electrons and formed new bonds. Life had to adapt or disappear.

The event that marked this transformation is known as the Great Oxygenation Event. It occurred between approximately 2.4 and 2.0 billion years ago and stands as one of the most significant turning points in Earth's history. The exact timing is difficult to pinpoint because the transition was not uniform across the globe, and oxygen levels fluctuated before stabilizing at higher concentrations. For countless early organisms, oxygen was toxic. It generated reactive molecules inside their cells, molecules that attacked proteins, damaged genetic material, and tore apart the membranes that held cellular structures together. These reactive oxygen species, as they are called, were lethal to organisms that had no defenses against them. Many of these organisms died out entirely. Others retreated to environments where oxygen couldn't reach them, hidden in deep sediments, isolated pools, or the guts of larger organisms that would evolve much later. Some of these anaerobic organisms still exist today, living in oxygen-free niches where they continue to practice the ancient chemistry of a vanished world.

Others adapted, and this adaptation was perhaps the most important development in the history of life. Through mutation and selection over vast stretches of time, some organisms developed molecules that could neutralize reactive oxygen species. They created enzymes that dismantled dangerous oxygen compounds before they could do harm. Even more significantly, some organisms began to use oxygen itself as a tool. They discovered that oxygen could be used to extract energy from food molecules with far greater efficiency than any process available before. By using oxygen to break down sugars and other compounds completely, these organisms could release more than ten times the energy compared to older methods. This process, called aerobic respiration, became one of the fundamental engines of life. Oxygen became a resource, a source of energy far more efficient than anything used by earlier life.

As oxygen accumulated in the atmosphere, rising to perhaps one percent or more of its current level, it also began to form a protective layer high above the surface. Sunlight in the ultraviolet range struck oxygen molecules in the upper atmosphere, splitting them apart into individual oxygen atoms. These free atoms were highly reactive and quickly recombined, but not always in the same way. Some bonded with other oxygen molecules to form ozone, a molecule made of three oxygen atoms instead of the usual two. Ozone has a unique property. It absorbs ultraviolet radiation very effectively, preventing much of this harmful energy from reaching the surface. This ozone layer, concentrated in what we now call the stratosphere, absorbed harmful ultraviolet radiation from the Sun, creating a safer environment at the surface. Before its formation, ultraviolet light would have been intense enough to damage the genetic material of any organism exposed to it for long periods. Life was largely confined to water, which provided some shielding, or to locations shaded from direct sunlight. Without this protection, life couldn't have expanded beyond the oceans or ventured into brightly lit environments. With it, the world became more welcoming. The rise of oxygen allowed life to grow in complexity and move into new environments that had previously been too hostile.

The presence of oxygen changed the oceans as well, and these changes were as significant as those on land. Iron began to settle out of the water as iron oxide, leaving the oceans clearer and fundamentally altering their chemistry. Waters that had once been murky with dissolved metals became more transparent, allowing sunlight to penetrate deeper and reach photosynthetic organisms in previously dark regions. The removal of iron also reduced the toxicity of seawater to certain forms of life, opening new ecological opportunities. New ecosystems formed as organisms evolved to use oxygen in their metabolism. These organisms could release far more energy from the same amount of food, allowing them to grow larger, move more actively, and develop more complex structures. The efficiency of aerobic respiration meant that cells could afford to invest energy in building intricate internal machinery, maintaining larger bodies, and supporting faster metabolisms.

The jump from simple cells to more complex organisms depended directly on this new flow of energy. Simple bacterial cells could survive on the modest energy yields of anaerobic processes, but larger, more complex cells required far more power. Around two billion years ago, a pivotal event occurred. Some bacteria that had learned to use oxygen were engulfed by other cells, and instead of being digested, they formed a partnership. These oxygen-using bacteria became mitochondria, the powerhouses of modern complex cells. This merger created eukaryotic cells, cells with internal compartments and far greater capabilities than their simpler predecessors. Oxygen made it possible for cells to cooperate, specialize, and form multicellular life. Without the energy provided by oxygen, the evolution of plants, animals, and fungi would not have been possible.

As oxygen levels rose further over hundreds of millions of years, Earth entered a new phase of its existence. The atmosphere became more stable in its composition, settling into patterns that could support more diverse forms of life. The range of possible climates expanded as oxygen interacted with other atmospheric components to moderate temperature and influence the distribution of heat around the planet. Weather systems changed. The surface environment grew more dynamic, supporting new forms of life that could tolerate or even require oxygen. Chemical weathering of rocks intensified as oxygen reacted with minerals exposed at the surface, releasing nutrients into rivers and eventually into the oceans. These nutrients fueled further growth of photosynthetic organisms, creating feedback loops that shaped the planet's chemistry. The foundations were laid for the eventual rise of plants, animals, and the ecosystems that shape the world today.

The rise of oxygen is a reminder that the most important transformations in Earth's history didn't come from sudden catastrophes or dramatic events, but from quiet changes that unfolded over vast stretches of time. Simple organisms changed the chemistry of the planet through countless small reactions, each one insignificant on its own but cumulative in effect. They reshaped the atmosphere and prepared the way for more complex life without any intention or awareness of what they were accomplishing. Without their work, Earth would have remained a world of simple cells living in a narrow range of conditions, forever limited by the constraints of anaerobic chemistry. The transformation took hundreds of millions of years, a timescale almost impossible for human minds to fully grasp, yet every stage was necessary.

Understanding the rise of oxygen shows how deeply life and the environment are connected. The atmosphere we breathe exists because early organisms altered it through their metabolic processes. The oxygen that makes up twenty-one percent of our air today is not a leftover from the planet's formation. It is a biological product, constantly replenished by photosynthetic organisms. The living world and the physical world evolved together, each shaping the other in an intricate dance of chemistry and biology. Life changed the planet by introducing a reactive gas that reshaped geology, climate, and the possibilities for future evolution. The planet, in turn, changed the course of life by creating new selective pressures and new opportunities. This partnership between biology and geology continues today, influencing everything from climate patterns to the cycling of nutrients through ecosystems. We live in a world built by bacteria, maintained by plants, and fundamentally dependent on processes that began billions of years ago.

The rise of oxygen was not the end of this story. It was the beginning of a new chapter that led to the evolution of complex life and all the diversity we see around us. With oxygen in the atmosphere and ozone in the sky protecting the surface from radiation, the stage was set for organisms to grow larger, more diverse, and more capable. The next steps in evolution would bring new forms adapted to use oxygen efficiently, new behaviors enabled by the energy it provided, and new possibilities for interaction and cooperation between organisms. Life was ready to reach toward complexity, building on the foundation laid by billions of years of microbial activity. The path toward animals, plants, and eventually human beings had begun, made possible by the slow accumulation of a gas that was once a poison and became the key to everything that followed.

Chapter Thirteen
Rise of Complex Life

For a long time, life on Earth remained small and simple. The first cells drifted through the oceans or clung to rocks, carrying out basic tasks that allowed them to survive. They copied themselves by dividing their genetic material and splitting in two. They reacted to their surroundings by sensing chemical gradients in the water and moving toward food or away from danger. They used energy in simple ways, breaking down molecules from their environment or capturing light from the Sun to fuel their internal processes. These early organisms were successful because they didn't need to be anything more than what they were. Earth was a planet filled with opportunities, and simple life could thrive with the tools it had. The oceans provided dissolved nutrients. Sunlight penetrated the upper layers of water. Chemical reactions occurred naturally on mineral surfaces. In this abundant world, a single cell with the right mechanisms could reproduce, spread, and persist for millions of years. But over time, small changes accumulated, and some cells began to work together. This cooperation marked the beginning of a new chapter in the story of life, one that would eventually lead to the rise of complex organisms.

The first major change came when cells learned to join in small groups and share responsibilities. Instead of living alone, they formed colonies. These colonies were not fully coordinated at first. Each cell still acted largely on its own, but the arrangement had benefits that natural selection could recognize and preserve. Cells in the center were protected from harsh conditions by the outer layers, shielded from waves, temperature fluctuations, and damaging ultraviolet radiation. Cells on the outside had better access to light or nutrients, positioning themselves where resources were most abundant. Over time, these cooperative groups became more stable. Chemical signals began to pass between neighboring cells, creating primitive forms of communication. Some cells became better at certain tasks and allowed others to focus on different roles. A cell that specialized in gathering nutrients could share its harvest with cells focused on reproduction. A cell that produced protective compounds could shelter the entire colony. Life was starting to explore the idea of specialization, and this exploration would prove to be one of the most important innovations in evolutionary history.

Eventually, some groups of cells took a larger step. Instead of acting as individuals that happened to live near each other, the cells in these groups began to function as a single organism. They stayed attached after dividing, maintaining physical connections through shared membranes and structural proteins. They coordinated their activities through chemical signaling networks that allowed one part of the organism to respond to what another part sensed. Some cells focused on movement, developing structures that could propel the organism through water. Others handled food processing, concentrating enzymes and absorbing nutrients more efficiently than any single cell working alone. Still others helped the organism maintain its shape, forming rigid frameworks or flexible boundaries that gave the whole structure integrity. These early multicellular organisms were simple by modern standards. They lacked organs, complex tissues, or sophisticated nervous systems. But they represented a major leap forward. Life could now grow larger, reaching sizes that single cells could never achieve without collapsing under their own metabolic demands. It could sense its environment more clearly, using different cells to detect different signals and integrating that information to make better decisions. It could respond with more flexibility than any single cell could manage, adapting its shape, movement, and behavior in ways that opened entirely new ecological niches.

As multicellular life grew more advanced, new forms of energy use opened possibilities that had never existed before. Organisms with photosynthetic cells gathered sunlight and grew in size and number, forming vast populations that filled the sunlit zones of the ocean with oxygen as a byproduct of their metabolism. This oxygen, once a rare gas in Earth's atmosphere, gradually accumulated to levels that would support entirely new types of life. Organisms that fed on other living things gained access to concentrated sources of energy stored in the complex molecules of their prey. These interactions created a delicate balance that pushed evolution forward in unexpected directions. When one group of organisms changed, others had to adapt or face extinction. If a new predator appeared with better vision or faster swimming, prey evolved ways to escape through camouflage, protective shells, or sudden bursts of speed. If a new food source became available, whether living or dead, organisms evolved ways to reach it through specialized feeding structures, digestive enzymes, or cooperative hunting behaviors. The environment and the organisms within it shaped one another through this constant cycle of change, creating feedback loops that accelerated the pace of evolution.

The Earth itself was also changing in ways that profoundly influenced the direction life would take. Oceans shifted as tectonic forces moved the planet's crust. Continents drifted across the surface, opening new seaways and closing others. Climates warmed and cooled in response to changes in ocean currents, volcanic activity, and the composition of the atmosphere. These changes created new environments and new challenges that life had never encountered before. Shallow seas formed along continental margins, creating zones where light penetrated to the bottom and nutrients concentrated in rich sediments. Deep ocean trenches opened, offering isolated habitats with extreme conditions. Coastlines expanded and contracted with sea level changes, alternately connecting and separating populations. Life responded with new strategies. Some organisms found ways to swim actively through the water column, developing fins, tails, and streamlined bodies that reduced drag. Others learned to anchor themselves in place, filtering food from passing currents or waiting for prey to drift within reach. Some grew hard shells for protection, precipitating minerals from seawater to build defensive armor. Others developed simple nervous systems that allowed them to respond quickly to danger, coordinating muscle contractions to flee or defend themselves with unprecedented speed. Over time, these incremental improvements created an explosion of diversity. Life spread into nearly every corner of the oceans, filling them with shapes and behaviors that had never appeared before. Competition intensified. Predators became more efficient. Prey became more elusive. And through it all, complexity increased.

Eventually, some organisms developed bodies with clear structures that suggested the basic body plans we recognize today. They had heads and tails, establishing a front end where sensory organs could gather information and a back end that provided propulsion. They had segments that allowed them to move more efficiently, with each section capable of independent motion that could be coordinated for swimming, crawling, or burrowing. They had muscles that could contract with greater strength, generating forces that enabled rapid movement and powerful actions. Their eyes detected light and shadows, using simple lenses or clusters of light sensitive cells to form images of their surroundings. Their nervous systems guided their actions, connecting sensors to muscles through networks of cells that transmitted signals with remarkable speed. These features allowed them to explore their environment in entirely new ways. They could hunt actively rather than waiting for food to arrive. They could navigate toward favorable conditions and away from threats. They could recognize other members of their species and coordinate behaviors for mating or defense. The oceans became crowded with life forms competing, cooperating, and adapting with increasing complexity. This was not a smooth or steady process. Mass extinctions occasionally reset the playing field, wiping out dominant groups and creating opportunities for survivors to diversify. But each time life recovered, it returned with new innovations built on the foundation of what had come before.

Some of the building blocks of consciousness didn't appear only in humans. Many animals experience smaller versions of awareness that reflect simpler brains. Dogs feel fear, excitement, and attachment. Crows use tools, plan ahead, and remember the faces of people who treated them badly. Octopuses solve puzzles and show individual personalities. These forms of awareness are not the same as human consciousness, but they show that conscious experience exists along a spectrum rather than as a sudden jump. Each species builds on the abilities of the one before it, adding new layers of memory, emotion, learning, and decision making. Understanding this continuum helps us see that human consciousness didn't appear out of nowhere. It has deep roots stretching back through millions of years of evolution.

This period of growth set the stage for the greatest transformation in the history of life. Once complex organisms filled the oceans, exploring nearly every available habitat beneath the waves, some began to explore shallow waters and muddy shores where the boundary between water and land blurred. They developed limbs strong enough to support them against the pull of gravity, transforming fins into sturdy appendages capable of bearing weight. Lungs adapted to breathe air, extracting oxygen directly from the atmosphere rather than filtering it from water. Skin changed to protect them from the Sun's unfiltered radiation and prevent their bodies from drying out in the open air. These creatures were the ancestors of the plants, animals, and insects that would one day spread across the land, building ecosystems as complex and diverse as those in the oceans they left behind.

The rise of complex life shows how powerful simple beginnings can be. Life started as small cells drifting in the water, invisible to the naked eye and barely distinguishable from the chemical reactions happening around them. But through countless generations of change and adaptation, through mutations that occasionally proved beneficial and combinations that sometimes created unexpected advantages, it built creatures capable of movement, sensing, feeding, and exploring. Every complex organism that exists today, including humans, can trace its lineage back to those early experiments in cooperation and structure. Evolution didn't follow a plan or aim toward a goal. It followed possibilities. Each new form of life opened the door to more complexity, and the planet became a canvas where life could paint its endless variations. Some experiments failed and left no descendants. Others succeeded spectacularly, founding lineages that would persist for hundreds of millions of years. The process was blind, driven by nothing more than reproduction, variation, and the simple fact that some variants survived better than others in the conditions they encountered.

As life continued to evolve, organisms became more capable, more aware of their surroundings, and more connected to the ecosystems they inhabited. They developed sophisticated sensory systems that could detect subtle changes in their environment. They evolved complex behaviors that allowed them to find food, avoid predators, and locate mates with increasing efficiency. They formed relationships with other species, sometimes competitive, sometimes cooperative, that shaped the structure of entire ecosystems. The next chapters in this story will explore how these complex life forms shaped the world, how they developed into creatures with remarkable abilities, and how they eventually created bodies as intricate as our own, with specialized organs, integrated systems, and the capacity for experiences we are only beginning to understand.

Chapter Fourteen
The Human Body

The human body is one of the most complex structures the universe has ever created. It contains roughly thirty-seven trillion cells, each performing tasks so intricate that they couldn't exist without billions of years of evolution guiding them into place. To put this in perspective, there are more cells in a single human body than there are stars in the Milky Way galaxy. Yet the body didn't appear suddenly or fully formed. It is the result of a long history that reaches back to the first atoms forged in the hearts of stars, the first molecules that assembled in ancient seas, and the first living cells that learned to harvest energy from their surroundings. To understand the human body, we must build it from the bottom up, starting with the smallest pieces and moving toward the remarkable whole.

Every part of the body begins with atoms. The carbon in your muscles was created when massive stars fused lighter elements under immense pressure and heat. The oxygen in your blood came from the same stellar furnaces. The calcium in your bones and the iron that carries oxygen through your bloodstream were born in the violent explosions of supernovae, where dying stars scattered their elemental wealth across space. These atoms drifted through the cosmic void for millions of years before gravity drew them into clouds of gas and dust that eventually collapsed to form the Sun and Earth. Over time, as Earth cooled and its surface solidified, these atoms became part of the oceans that covered the young planet, the rocks that formed its crust, and the atmosphere that wrapped around it like a protective shell. When life first appeared in Earth's waters, it gathered these wandering atoms into molecules, and molecules organized themselves into the earliest cells. The same atoms that once floated freely through space now make the structures that allow you to think, move, breathe, and feel. Your body is made of stardust, not as poetry but as scientific fact.

Inside your body, atoms join together to form molecules that work together in highly organized ways. Water makes up about sixty percent of your body weight and forms the liquid environment where all your cells operate. Every chemical reaction in your body takes place in water or involves water molecules. Proteins are the workhorses of your cells. They are long chains of smaller units called amino acids, folded into precise three-dimensional shapes that determine their function. Some proteins form the physical structures of your cells and tissues, creating scaffolding and support. Others act as enzymes that speed up chemical reactions, allowing processes that would take years to happen in milliseconds. Still others transport molecules from one place to another or receive signals from outside the cell and trigger responses inside. Lipids, which include fats and oils, create the membranes that surround your cells and define their boundaries. These membranes are not solid barriers but flexible sheets that constantly shift and change, allowing certain molecules to pass through while keeping others out. Carbohydrates provide quick energy when broken down and also attach to proteins and lipids on cell surfaces, creating identity markers that help cells recognize one another. DNA, the molecule of heredity, stores the instructions that allow your body to grow, repair itself, and create new cells. It exists as a double helix, a twisted ladder where each rung contains coded information. All of these molecules interact in precise patterns, guided by the laws of chemistry and physics, forming the foundation upon which life is built.

Cells are the next level of organization, and they represent something extraordinary in the universe: matter that has learned to maintain itself, to grow, and to reproduce. A cell is like a small city with specialized structures inside that carry out essential tasks. The nucleus serves as the control center, housing DNA and directing the cell's activities.

Mitochondria act as power plants, breaking down food molecules and releasing energy in a form the cell can use. These tiny structures were once independent bacteria that formed a partnership with other cells billions of years ago, and they still carry their own DNA separate from the nucleus. The endoplasmic reticulum and Golgi apparatus work like factories and shipping centers, creating proteins and packaging them for delivery to other parts of the cell or to destinations beyond the cell membrane. Ribosomes read the instructions in messenger RNA molecules and assemble proteins one amino acid at a time. Lysosomes contain powerful enzymes that break down worn-out cell parts and digest foreign material. Each cell is surrounded by a membrane made of lipids and proteins that acts as a selective barrier, allowing nutrients and oxygen to enter while keeping out harmful substances and letting waste products exit. Cells divide when they grow too large or when the body needs more of them, copying their DNA and splitting into two daughter cells. They repair damage to themselves using molecular machinery that detects and fixes errors. They communicate with one another through chemical signals, receiving messages from neighboring cells and from distant parts of the body. These signals tell cells when to grow, when to stop growing, when to move, and when to die. Cells work together to create tissues, and tissues join to form organs, each level of organization adding new capabilities that didn't exist at the level below.

Organs are collections of different tissue types that specialize in certain tasks. The heart is a muscular pump that contracts rhythmically, sending blood through your body in a circuit that never stops from before you are born until the moment you die. It beats about one hundred thousand times each day, moving roughly two thousand gallons of blood. The lungs bring in oxygen from the air and release carbon dioxide, a waste product of cellular respiration. They contain millions of tiny air sacs called alveoli where oxygen passes into the bloodstream and carbon dioxide passes out. The surface area of all these alveoli combined would cover nearly half a tennis court if spread flat. The liver is a chemical processing plant that performs over five hundred different functions. It filters toxins from your blood, stores energy in the form of glycogen, produces proteins needed for blood clotting, breaks down old blood cells, and helps digest fats. The kidneys clean your blood by filtering out waste and excess water, producing urine that carries these substances out of your body. They also regulate the balance of salts and minerals in your blood, keeping conditions stable even as your diet and activity levels change. The stomach and intestines form a long tube where food is broken down mechanically and chemically, then absorbed into the bloodstream. The small intestine alone is about twenty feet long and lined with millions of finger-like projections called villi that increase its surface area for absorption. Each organ is made of many types of cells working together, each cell contributing its specialized function to the organ's larger purpose. And each organ supports the others in an interconnected system that keeps the whole body functioning as one integrated unit.

The nervous system is one of the most remarkable parts of the human body. It connects every part of you and allows you to sense the world, respond to it, and exist as a conscious being. Nerve cells, called neurons, are unlike any other cells in your body. They send electrical and chemical signals through long extensions called axons that can stretch for several feet, reaching from your spinal cord to the tips of your toes. These signals move quickly, traveling at speeds up to two hundred miles per hour in some neurons, allowing you to react to your environment almost instantly. When you touch something hot, neurons in your skin detect the heat and send a signal up your arm, into your spinal cord, and to your brain. The brain processes this information and sends signals back down to muscles in your hand, telling them to pull away. This entire sequence happens in a fraction of a second. Neurons communicate with each other at junctions called synapses, where the electrical signal in one neuron triggers the release of chemical messengers that cross a tiny gap and bind to receptors on the next neuron, continuing the signal. Touch, sound, taste, smell, and sight all depend on specialized neurons that convert physical stimuli into electrical signals. Neurons in your skin respond to pressure. Neurons in your ears respond to vibrations. Neurons in your eyes respond to light. These signals travel through networks of neurons to your brain, which interprets them and creates your experience of the world. The brain itself contains roughly eighty-six billion neurons, each connected to thousands of others, forming a network so complex that it gives rise to thought, memory, emotion, and consciousness itself.

The circulatory system carries nutrients, oxygen, and chemical signals to every cell in your body. The heart pumps blood through a vast network of tubes: arteries that carry blood away from the heart, veins that carry blood back to the heart, and capillaries so small that they can only be seen under a microscope but so numerous that they reach every tissue. If you could lay out all the blood vessels in your body end to end, they would stretch for about sixty thousand miles, enough to circle the Earth more than twice. Blood is not simply a liquid but a complex tissue made of different components suspended in plasma, a watery fluid that contains dissolved proteins, salts, and other substances. Red blood cells are packed with a protein called hemoglobin, which contains iron atoms that bind to oxygen molecules in your lungs and release them in tissues throughout your body. Each red blood cell lives for about four months, during which it makes roughly a quarter of a million trips around your circulatory system. White blood cells patrol your body looking for signs of infection or damage. There are several types of white blood cells, each with its own strategy for defending you. Some engulf and digest invaders. Others produce antibodies. Still others remember past infections. Platelets are cell fragments that help repair injuries by clinging to damaged blood vessel walls and forming clots that stop bleeding. Blood delivers everything cells need to survive and removes carbon dioxide and other waste products, carrying them to the lungs and kidneys where they can be expelled from the body. This constant circulation keeps conditions stable inside your body even as the external environment changes.

The skeletal system provides structure and protection. Bones are not solid, lifeless objects like the stones they resemble. They are living tissues filled with cells that repair and rebuild themselves constantly. The outer layer of bone is dense and hard, made of calcium and phosphorus arranged in a crystalline structure that gives bone its strength. Inside, bone is more porous, with a spongy structure that makes bones lighter without sacrificing much strength. At the very center of many bones is the marrow, where blood cells are born. Your skeleton contains two hundred and six bones in adulthood, each shaped precisely for its role. Long bones in your legs act as pillars that support your weight. Flat bones in your skull protect your brain. Bones store minerals like calcium and release them into the bloodstream when the body needs them. Muscles attach to bones through tough bands of tissue called tendons. When muscles contract, they shorten and pull on the bones they are attached to, creating movement. Muscles work in pairs: when one contracts and pulls a bone in one direction, another muscle on the opposite side relaxes, and then reverses the motion when needed. Skeletal muscles are under your conscious control, allowing you to walk, lift objects, speak, and perform every voluntary action your body is capable of. But there are also involuntary muscles in your organs that work without your awareness, pushing food through your digestive system and controlling the diameter of your blood vessels.

The immune system defends the body from harmful invaders with a sophistication that rivals any military defense network. It patrols your body constantly, distinguishing between what belongs and what does not. White blood cells circulate through your bloodstream and lymphatic system, a parallel network of vessels that drains fluid from your tissues. Some immune cells act as scouts, sampling molecules from their surroundings and presenting them to other immune cells. If they find something foreign, such as proteins from bacteria or viruses, they trigger an immune response. Other immune cells attack directly, engulfing invaders and destroying them with toxic chemicals. Still others produce antibodies, Y-shaped proteins that bind to specific molecules on the surface of pathogens and mark them for destruction. The immune system recognizes millions of different threats with remarkable specificity. It can tell the difference between different strains of the same virus. It can distinguish your own cells from nearly identical cells from another person. Perhaps most remarkably, the immune system remembers. After you fight off an infection, some immune cells remain in your body for years or even decades, ready to respond faster and more effectively if the same pathogen returns. This immunological memory is what makes vaccines possible and why you usually get certain illnesses only once. The immune system is constantly active, even when you feel perfectly healthy, quietly eliminating damaged cells and fighting off pathogens before they can cause disease.

All of these systems work together in a harmony so smooth that most of it happens without your awareness. Your body maintains a state called homeostasis, keeping internal conditions stable even as the external environment changes. When you get too hot, blood vessels in your skin dilate to release heat and sweat glands activate to cool you through evaporation. When you get too cold, blood vessels constrict to conserve heat and muscles shiver to generate warmth. When your blood sugar rises after a meal, your pancreas releases insulin to help cells absorb glucose and store it for later. When blood sugar falls, other hormones trigger the release of stored glucose back into the bloodstream. Your body keeps you alive every second of your life without asking for your attention. It repairs wounds by calling immune cells to the site of injury, forming clots to stop bleeding, and producing new tissue to fill the gap. It replaces old cells with new ones: your skin completely renews itself every few weeks, your red blood cells are replaced every few months, and even your bones are constantly being broken down and rebuilt. It fights infections you never know you had. It regulates your temperature, balances chemicals in your blood, adjusts your heart rate and breathing based on your activity level, and keeps every organ in motion. This coordination is possible because evolution shaped each system over an immense span of time, refining its structure and improving its function through countless generations. Each system evolved not in isolation but in relationship to all the others, creating a web of interactions so intricate that we are still discovering how it all works.

The human body is a masterpiece constructed from the simplest ingredients in the universe. It is built from atoms formed in the nuclear furnaces of stars and scattered through space by stellar explosions, arranged into molecules that took shape in ancient oceans when Earth was young, organized into cells that trace their ancestry back to the first living creatures that emerged more than three and a half billion years ago. Your body is a living record of the universe's history, carrying within it evidence of cosmic events that happened long before Earth existed. It carries the story of matter becoming life and life becoming aware of itself. When you look at your hand, you are seeing atoms that were once part of stars looking back at themselves across billions of years. Understanding the body from the bottom up reveals that we are not separate from the universe, not observers standing outside looking in, but participants, made of the same substance as planets and stars. We are one of the ways the universe has organized itself into something that can think, feel, wonder, and ask questions about its own existence. We are, in the most literal sense, the universe becoming conscious of itself. The human body, built from stardust and shaped by evolution, stands as one of the most intricate and remarkable expressions of what matter can become.

Chapter Fifteen
Evolution of Intelligence

Intelligence didn't appear suddenly. It didn't arrive fully formed or spring into existence in a single moment. It emerged slowly, shaped by the pressures of survival and the opportunities that life discovered over billions of years. The earliest forms of intelligence were simple, almost invisible compared to what we see in modern animals. They were not thoughts or plans. They were reactions. They were the ability of a cell to sense its environment and respond in a way that increased its chances of survival. This was the beginning of intelligence, and it started long before there were animals, brains, or anything that resembled conscious thought.

The first organisms couldn't think, but they could react. Single-celled bacteria, appearing more than three billion years ago, developed molecular sensors embedded in their membranes. These sensors could detect changes in temperature, light, or the presence of specific chemicals. When a bacterium encountered a harmful substance, chemical signals triggered changes in the proteins that controlled its flagellum, causing it to tumble and swim in a different direction. When it detected nutrients, different signals kept it moving forward. This process, called chemotaxis, was entirely mechanical. No awareness guided it. Yet it represented a fundamental breakthrough: the ability to gather information from the environment and use that information to influence behavior. These reactions were simple, yet they gave early life a small advantage. When one cell responded slightly better than another, it had a better chance of surviving and reproducing. Over time, these tiny advantages accumulated. Natural selection favored organisms that could sense their environment more accurately and respond more effectively. Intelligence began as a tool for survival, shaped by the demands of a changing world.

As life grew more complex, so did its systems for processing information. Multicellular organisms, which began to dominate the oceans around six hundred million years ago, faced new challenges. A single cell could coordinate its own functions, but a body made of millions or billions of cells needed a way to synchronize movement and share information across distances. The solution came through the evolution of nerve cells, specialized cells capable of transmitting electrical signals much faster than chemical diffusion alone could achieve. These nerve cells gathered into networks that could send signals quickly from one part of the organism to another. A simple nerve net, like those found in modern jellyfish and sea anemones, allowed an organism to contract its entire body when touched or move toward food when chemical signals were detected. Although primitive by modern standards, this was an important step. It marked the beginning of organized information processing, a foundation that would one day support memory, learning, and conscious awareness.

Over millions of years, nerve nets became more specialized and centralized. Random networks gave way to organized structures where some clusters of nerve cells performed specific tasks, such as coordinating rhythmic movement or detecting changes in light intensity. These clusters represented the first hints of a brain, a central processing unit that could integrate information from multiple sources. Animals that could move more effectively gained significant advantages in finding food and avoiding danger, and the ability to navigate through three-dimensional space encouraged the development of more sophisticated sensory organs. Eyes evolved independently in different lineages, beginning as simple patches of light-sensitive cells and eventually developing lenses, focusing mechanisms, and the neural circuitry to process visual information.

Ears allowed animals to detect vibrations and locate sounds in their environment. Chemical sensors became more refined, able to distinguish between subtle differences in taste and smell. The world became richer in detail, and organisms capable of interpreting that detail gained access to new opportunities for survival and reproduction.

The rise of predators and prey accelerated the evolution of intelligence in ways that transformed life on Earth. When some animals began to hunt others, an evolutionary arms race began. Predators needed to track movement, anticipate the behavior of their prey, and react quickly to sudden changes in direction. Prey needed to sense danger before it arrived, remember which areas were safe and which were dangerous, and find food while constantly monitoring for threats. This created a cycle of continuous improvement driven by competition. Survival required sharper senses, faster reactions, and better coordination between perception and movement. Over time, simple brains evolved into more complex structures with distinct regions dedicated to different functions. Some areas specialized in processing sensory input. Others coordinated motor control. Still others began to store information about past events. Memory appeared, first in simple forms that lasted only minutes or hours, then in more permanent forms that could persist for days, months, or even years. Learning appeared alongside memory, allowing animals to modify their behavior based on experience. Animals began to adapt within their own lifetimes, not just over generations. A fish that learned to associate a certain location with food could return to that location repeatedly. A mouse that encountered a predator in one area could avoid that area in the future. This flexibility gave animals an enormous advantage in unpredictable environments.

As brains grew larger and more intricate, animals developed behaviors that depended heavily on experience rather than pure instinct. They learned to navigate complex environments using landmarks and spatial memory. They recognized specific dangers and adjusted their behavior accordingly. They remembered productive food sources and returned to them when needed. Some species developed social behavior, living in groups that required communication, cooperation, and the ability to recognize individual members. Social living encouraged even more intelligence because it introduced new kinds of problems to solve. Individuals had to understand their place within a social hierarchy, recognize both allies and rivals, and respond appropriately to changing social conditions. They needed to interpret the intentions of others, predict their behavior, and coordinate their own actions with the group. This demanded a higher level of awareness and more advanced problem-solving abilities than solitary life required. Animals that could navigate social complexity gained access to the benefits of group living: shared vigilance against predators, cooperative hunting, communal care of young, and the transmission of learned behaviors from experienced individuals to novices.

The ancestors of mammals took this evolutionary trajectory further. Emerging after the great extinction that ended the age of dinosaurs, mammals developed large brains relative to their body size, with structures that supported sophisticated learning, emotional responses, and long-term memory. The mammalian brain included a well-developed cerebral cortex, a layered structure that would become the seat of higher cognitive functions. Mammals invested heavily in their offspring, providing extended parental care that created opportunities for learning. Young mammals could observe their parents, practice new skills in relative safety, and gradually master the complex behaviors needed for survival. This period of dependency meant that more information could be passed from one generation to the next through observation and teaching rather than through genetic inheritance alone. Mammals could form complex social bonds that lasted years or even lifetimes. These traits increased the importance of intelligence as a survival tool. Over millions of years, certain groups of mammals evolved especially large and flexible brains capable of sophisticated cognitive tasks. Among them were the primates, a diverse group that would eventually include monkeys, apes, and humans.

Primates evolved in forest environments where three-dimensional movement through trees created intense selection pressure for specific cognitive abilities. Depth perception became crucial for judging distances between branches. Hand-eye coordination allowed precise grasping and manipulation of objects. Social cooperation enabled group defense against predators and collaborative foraging. Their brains grew substantially to support these traits, with expanded visual cortices for processing detailed spatial information and enlarged frontal regions for planning and decision-making. Primates developed the ability to learn from one another through observation, a process that accelerated the spread of new behaviors within populations. They communicated through complex combinations of vocalizations, facial expressions, and body language. They formed long-lasting relationships built on reciprocity and recognition of past interactions. Problem-solving abilities advanced significantly, with some species demonstrating the capacity to use tools, recognize themselves in mirrors, and understand cause and effect relationships. Planning emerged, allowing primates to think ahead and modify their behavior in anticipation of future events. Each step in this progression required more brain tissue, more energy to fuel that tissue, and more time for young primates to develop their cognitive abilities. Each step also created new opportunities for intelligence to grow.

Tool use became one of the clearest signs that intelligence was shifting toward flexible problem-solving rather than fixed instinct. Some primates began using sticks to extract insects from logs, stones to crack open hard nuts, and leaves as makeshift sponges for drinking water. These behaviors didn't appear all at once. They spread through observation, imitation, and practice. A young primate watched an older one solve a problem and then repeated the behavior with increasing skill. Tool use allowed access to foods that were otherwise unreachable and created a new selection pressure favoring individuals who could learn, innovate, and teach. This expanded the cognitive demands of daily life, pushing brain evolution even further.

As brains grew larger and more capable, they also became more expensive. Neural tissue consumes enormous amounts of energy. In modern humans, the brain makes up only about two percent of total body weight, yet it uses roughly twenty percent of the body's energy at rest. This imbalance began long before humans existed. Early mammals and primates with larger brains had to eat more calorie-dense foods and rely on social cooperation to cover these energy costs. Bigger brains only evolved when they gave a survival advantage strong enough to outweigh their metabolic burden. This trade-off shaped the course of intelligence, limiting when and where large brains could evolve but rewarding species that could support them.

The evolution of intelligence reached a critical turning point when early humans appeared roughly two million years ago. Their ancestors already possessed strong social structures, the ability to walk upright on two legs, and hands freed from locomotion that were capable of fine manipulation and precise tool use. These abilities allowed early humans to create and use tools in ways no other species could match. Simple stone tools for cutting and scraping gave way to more sophisticated implements requiring planning and foresight to manufacture. Tools increased survival rates by making food more accessible and predators less dangerous, and better tools required better planning, greater manual dexterity, and more effective communication between individuals. Language began to take shape as early humans worked together on complex tasks, shared knowledge about resources and dangers, and taught one another new techniques. The evolution of language represented a profound transformation. Unlike the calls and gestures of other animals, human language could refer to things not immediately present. It could describe past events, discuss future possibilities, and convey abstract ideas. Language allowed successful strategies and hard-won knowledge to pass from one generation to the next without each individual having to rediscover everything through trial and error. This accelerated the spread and accumulation of knowledge in ways that had never before occurred in the history of life.

Once humans could think symbolically, imagine future scenarios, and communicate abstract concepts, intelligence transformed into something fundamentally new. It became a force that shaped not just individual survival, but entire environments and ecosystems. Humans learned to control fire, which provided warmth, protection from predators, and the ability to cook food, making more nutrients available and reducing the energy needed for digestion. They built shelters adapted to different climates, allowing them to expand into regions that would otherwise have been uninhabitable. They created art, suggesting an inner life rich with symbolic meaning and aesthetic appreciation. They formed increasingly complex communities with division of labor, trade networks, and shared cultural practices. The development of agriculture roughly ten thousand years ago marked another major transition. By deliberately cultivating plants and domesticating animals, humans could produce more food in a smaller area, which allowed stable, permanent settlements. Population densities increased. Specialized roles emerged. Complex societies developed with new forms of organization and cooperation. The creation of writing systems allowed ideas, laws, histories, and technical knowledge to exist beyond the limits of human memory. Information could be preserved, refined, and transmitted across vast distances and spans of time. Intelligence became cultural, not just biological. Knowledge spread through populations, accumulated across generations, and grew exponentially with each passing century.

The evolution of intelligence is a story of slow, incremental advancement shaped by countless challenges and opportunities. It began with simple chemical reactions in single-celled organisms and grew into the ability to reason abstractly, imagine distant futures, and reflect on existence itself. Intelligence is not separate from the natural world or exempt from its laws. It is the outcome of billions of years of life adapting to survive in changing environments. Every thought that crosses your mind, every idea you consider, and every moment of awareness you experience is supported by a biological structure that nature built through immense spans of time and countless generations of gradual modification. The human brain, with its eighty-six billion neurons and trillions of connections, represents the current pinnacle of this process, but it is not the end. Evolution continues, though its future direction remains uncertain.

Understanding the evolution of intelligence helps us see ourselves more clearly and more honestly. The human mind is extraordinary in its capabilities, but it didn't appear by chance, divine intervention, or sudden creation. It is the result of a long and continuous process that began with the simplest forms of life responding to their environment. Our ability to think, to wonder, to ask questions, and to understand the universe is the latest step in a journey that spans the entire history of life on Earth. In the next chapter, we explore how this developing intelligence eventually gave rise to consciousness itself, and how the brain became the structure that supports not just behavior and learning, but our inner subjective experience of being alive.

Chapter Sixteen

The Birth of Consciousness

The human brain is one of the most complex structures in the universe. Within the space of roughly three pounds of soft tissue reside approximately eighty-six billion neurons, each capable of forming thousands of connections with its neighbors. These neurons are made of ordinary atoms assembled into cells, and those cells form networks so intricate that no two brains are exactly alike. The total number of possible connections between neurons exceeds the number of atoms in the observable universe. This structure gives rise to thoughts, memories, emotions, and awareness. Yet the brain didn't appear fully formed. It is the result of a long chain of events that began with the first living cells on Earth nearly four billion years ago. To understand the brain, we must follow the same path we used to understand the body. We must build it from the bottom up and see how a physical organ made of matter can give rise to something as remarkable as consciousness.

The earliest forms of life had no nervous systems at all. Single-celled organisms and simple multicellular creatures moved and reacted based on direct chemical responses to their environment. A bacterium might swim toward nutrients by detecting chemical gradients in the water around it. These responses were automatic and immediate, determined entirely by the molecular machinery inside the cell. There was no processing, no decision, no awareness. The organism simply responded to chemical signals the way a ball responds to gravity. Yet even this simple form of responsiveness laid the groundwork for everything that would follow.

As organisms grew more complex and their bodies became larger, direct chemical responses became insufficient. A creature the size of a worm couldn't rely on chemicals slowly diffusing through its body to coordinate its movements. It needed something faster. Some organisms developed specialized cells that could carry signals through their bodies at much greater speeds. These early nerve cells, or neurons, used electrical impulses that traveled along their length, allowing information to move from one end of the body to the other in a fraction of a second. These signals allowed organisms to detect danger, find food, and coordinate movement far more effectively than their simpler ancestors. Over time, these cells formed networks that connected different parts of the body. In simple creatures like jellyfish and flatworms, these networks guided basic behaviors such as moving toward light or contracting away from threats. In more complex organisms, they began to form clusters that processed information more effectively, creating the first hint of centralized control.

As evolution continued, these clusters grew larger and more organized. They became the first primitive brains. These early brains, found in creatures like ancient fish and early amphibians, handled movement, feeding, and survival. They didn't think or plan in the way humans do, but they allowed animals to adapt to their surroundings with greater skill. Different environments pressured these brains to evolve in different directions. Creatures that hunted needed strong senses and quick reactions, leading to the development of enhanced visual systems and rapid motor control. Creatures that hid needed acute awareness of predators, which drove the evolution of sophisticated sensory processing. Creatures that lived in groups needed ways to communicate and recognize one another, adding layers of social processing to their neural architecture. All of these demands shaped the development of the nervous system and slowly pushed evolution toward greater complexity. Natural selection favored any innovation that improved an organism's ability to survive and reproduce, and many of those innovations involved improvements to the brain.

Eventually, some animals developed brains capable of learning. Instead of relying solely on instincts encoded in their genes, they could remember past experiences and use those memories to guide their actions. A bird that ate a brightly colored insect and became sick would remember the color and avoid similar insects in the future. This ability gave organisms a significant advantage. Learning allowed them to adapt in a single lifetime instead of waiting for genetic changes to spread over many generations. An animal that could learn from experience could survive in changing environments that would kill creatures relying only on fixed behaviors. As learning became more valuable, evolution favored brains that were larger and more flexible, with more neurons and more complex networks capable of storing and retrieving information. The ancestors of mammals began to show these traits hundreds of millions of years ago, and over time their brains grew dramatically in both size and complexity.

The human brain is built from this long history, and its structure reflects every stage of this evolutionary journey. Deep within the brain lies the brainstem, the most ancient part, which controls fundamental survival functions like breathing, heartbeat, and sleep cycles. These structures resemble those found in reptiles and even fish, a reminder of our deep evolutionary roots. Surrounding the brainstem is the limbic system, which handles emotions, motivation, and basic memory formation. This region became more elaborate in early mammals and plays a crucial role in forming emotional responses and social bonds. Above and around these ancient structures lies the cerebral cortex, the wrinkled outer layer that is responsible for thought, memory, reasoning, and imagination. This layer, particularly the prefrontal cortex at the front of the brain, is what allows humans to plan ahead, understand abstract ideas, reflect on their own existence, and imagine possibilities that don't yet exist. The cortex is vastly expanded in humans compared to other animals, giving us capabilities that no other species possesses.

Humans also developed something called meta-awareness, which means being aware of your own awareness. It is the ability to notice your thoughts while you are thinking them, to step back and observe your own reactions, fears, memories, and intentions as if you were watching yourself from the outside. Meta-awareness allows you to recognize when you are upset, understand why you made a mistake, or deliberately change your behavior instead of acting on impulse. It is also what makes reflection, imagination, and long-term planning possible. Animals may feel emotions and make decisions, but humans can examine those emotions, question their decisions, and reshape them. This extra layer of awareness is one of the key differences between human consciousness and the consciousness found in other animals.

The brain works because its cells, called neurons, communicate through tiny electrical pulses and chemical signals. Each neuron is a living cell with a body, branching extensions called dendrites that receive signals, and a long fiber called an axon that sends signals to other neurons. When a neuron is sufficiently stimulated, it generates an electrical impulse that races down its axon at speeds up to several hundred miles per hour. At the end of the axon, the signal reaches a junction called a synapse, where the neuron releases chemical messengers called neurotransmitters. These molecules drift across a microscopic gap and bind to receptors on the next neuron, either encouraging or discouraging it from firing its own signal. Each neuron connects to thousands of others through these synapses, forming networks of staggering complexity that can change and strengthen over time. These networks allow the brain to store memories, solve problems, and learn new skills. When you think, countless signals move through your brain in patterns shaped by your past experiences. These patterns shift and reorganize every time you learn or remember something. The connections between neurons that are used frequently grow stronger, while those that are rarely used weaken and may eventually disappear.

The brain is constantly changing and reshaping itself, even when you are at rest or asleep.

Consciousness arises from the activity of these networks. It is the experience of being aware, of having thoughts and feelings, and of recognizing yourself as an individual distinct from the world around you. Scientists still don't fully understand how consciousness emerges from matter, and it remains one of the deepest mysteries in science. But they know that it depends on the communication between different regions of the brain working together in synchronized patterns. When sensory information from your eyes and ears combines with memories from your past, emotions from your limbic system, and planning from your prefrontal cortex, a unified sense of self appears. You become aware not just of your surroundings but of the fact that you are aware. You can think about your own thoughts, remember your own memories, and imagine your own future. This awareness is one of the most remarkable properties of the universe, created not from magic or any substance beyond the physical world, but from the interactions of billions of neurons following the same laws of nature that govern everything else.

The brain is also profoundly shaped by the world around it. Every experience changes the connections between neurons. When you learn something new, specific synapses strengthen, creating physical traces of that knowledge in the structure of your brain. Memories form when these connections strengthen and stabilize, creating patterns that can be reactivated later. The more you recall a memory, the stronger those connections become. Emotions influence how memories are stored, which is why you remember emotionally significant events more vividly than mundane ones. The emotional centers of the brain add a tag of importance to certain memories, ensuring they are preserved more robustly. Language, culture, and relationships all play crucial roles in shaping how the brain develops and how consciousness expresses itself. The language you speak literally changes the structure of your brain, creating networks specialized for the sounds and grammar of that language. The culture you grow up in influences which concepts you find natural or strange, which emotions you recognize and express, and how you understand yourself in relation to others. No two people see the world in the same way because no two brains have experienced life in the same way. Your mind is a unique reflection of your history, your environment, and the particular structure of your brain as it has been shaped by both your genes and your experiences.

Even with everything science has discovered about the brain, one mystery remains unresolved. Researchers can explain how neurons fire, how memories form, and how the brain organizes information, but they cannot yet explain why those physical processes create subjective experience. This challenge is sometimes called the hard problem of consciousness. It asks why electrical signals and chemical exchanges in the brain produce feelings, thoughts, dreams, and the sense of being "you" instead of functioning like a complex machine without inner experience. Scientists can describe the mechanics of the brain in great detail, but the reason those mechanics produce awareness remains an open question.

The birth of consciousness marks one of the most profound transitions in the story of the universe. For billions of years, matter existed without awareness. Stars formed and died. Galaxies spun through space. Chemical reactions produced increasingly complex molecules. But none of it was witnessed. There was no experience, no understanding, no wonder. Then, through the long process of evolution, matter organized itself into forms capable of sensing, remembering, and eventually reflecting. After billions of years of physical and biological evolution, the universe produced beings capable of understanding it. The same atoms that formed in the hearts of dying stars billions of years ago, the same atoms that drifted through space and gathered into planets, now form the neurons in your brain. These atoms now think, feel, and question their own origins. Consciousness allows the universe to look back at itself through human eyes and wonder how it came to be. In a very real sense, we are the universe becoming aware of itself.

As we continue, we will turn our attention from the human mind to the larger structures that surround us. The Earth, the planets, and the wide universe all play roles in the story of existence.

Chapter Seventeen
Formation of the Solar System

Long before Earth formed, long before the planets settled into their familiar paths around the Sun, the solar system existed only as a cloud of gas and dust drifting through the Milky Way. This cloud, known to astronomers as the solar nebula, spanned billions of kilometers across the empty reaches of space. It contained the remnants of earlier stars that had lived and died long before our Sun, their final explosions scattering heavy elements into the interstellar medium. Inside this vast accumulation were the elements that would one day form oceans, mountains, planets, and living creatures. Hydrogen and helium dominated its composition, making up roughly ninety-eight percent of the mass, but crucial heavier elements were there too. Carbon, oxygen, nitrogen, iron, silicon, and dozens of other elements floated among the lighter gases, the gift of ancient supernovae. The story of the solar system begins with this cloud, shaped by gravity, turbulence, and time. What we know as home began as a scattered mixture floating in the quiet spaces between the stars, cold and dark, with temperatures hovering just a few degrees above absolute zero.

Something disturbed this cloud approximately four and a half billion years ago. It may have been the shock wave of a nearby exploding star, a supernova whose expanding shell of gas slammed into the nebula and compressed its edges. Or perhaps the passing of a massive object through the galaxy triggered ripples of density that propagated through the cloud's structure. Whatever the cause, a critical threshold was reached. The cloud began to collapse inward under its own gravity. Once this process started, it accelerated. As the cloud shrank, its rotation quickened, a consequence of the conservation of angular momentum. This principle governs everything from spinning figure skaters to collapsing stellar nurseries. Just as a skater spins faster when pulling in their arms, the contracting cloud spun faster as its radius decreased. This rotation prevented the cloud from collapsing uniformly in all directions. Instead, it shaped the collapsing material into a wide, flat disk with a bulging center, a structure called a protoplanetary disk. The same physics that creates this shape also explains why galaxies form spirals and why Saturn has rings. Most of the material gathered at the center where gravity was strongest, and this region grew hotter and denser as more matter fell inward, compressed by the weight of everything above it.

All planets eventually formed in nearly the same plane and moved in the same direction around the Sun because the collapsing cloud forced everything into a spinning, flattened disk. Gas and dust that tried to move in any other direction were dragged into the disk by friction and collisions. Random motions were erased by constant interactions until almost all material followed the same rotation. This process locked the solar system into a single orientation, which is why the planets orbit in orderly paths rather than swirling in every direction like debris from an explosion.

At the core of the disk, temperatures climbed steadily as gravitational energy converted into heat. The center became a furnace, reaching millions of degrees. When the core temperature exceeded ten million degrees Kelvin, a threshold was crossed. Hydrogen atoms, stripped of their electrons by the intense heat and moving at tremendous speeds, collided with such force that they overcame the electromagnetic repulsion between their positively charged nuclei. They fused together, combining four hydrogen nuclei into one helium nucleus through a series of nuclear reactions. This fusion released energy according to Einstein's equation, converting a small amount of mass directly into light and heat. The energy poured outward, creating pressure that balanced the inward pull of gravity. This marked the birth of the Sun.

It became a new star, burning hydrogen in its core and surrounded by a hot, swirling disk of leftover gas and dust. The Sun consumed most of the original cloud, claiming more than ninety-nine percent of the available mass. But the remaining material didn't disappear. It continued to orbit the young star in the flattened disk, slowly cooling as it radiated heat into space and separating into distinct regions that would form the planets.

The Sun kept almost all the mass because gravity becomes overwhelmingly dominant at large scales. As the cloud collapsed, any region even slightly denser than the rest pulled in more material, and the center quickly became that region. Every piece of dust or gas that drifted inward was accelerated toward the growing star, heating the core and increasing its pull. Planets formed only from the leftover material that never reached the center, which is why they contain just a tiny fraction of the solar system's total mass.

Closer to the Sun, within a few astronomical units of the newborn star, the heat was too intense for ice or light compounds to remain solid. Water, methane, and ammonia would evaporate immediately in this zone. Only substances with high melting points could survive as solids. Metals like iron and nickel, and silicate minerals containing silicon and oxygen, remained stable in solid form despite the fierce temperatures. The inner region of the disk therefore became the birthplace of rocky planets, worlds built from refractory materials that could withstand the heat. Farther out, beyond roughly three to five astronomical units from the Sun, the temperature dropped below the freezing point of water. This boundary, called the frost line or snow line, marked a dramatic change in the disk's composition. Beyond it, ices could condense and gather into larger bodies alongside the rocky materials. Water ice, methane ice, and ammonia ice swelled the available solid matter by several times, providing far more building material than existed in the inner system. These differences in temperature and available materials shaped the fundamental architecture of the solar system. The inner planets of Mercury, Venus, Earth, and Mars formed from silicate minerals and metals, remaining relatively small because they had limited raw materials. The giant planets of Jupiter and Saturn formed farther away where ice, gas, and lighter materials could accumulate in abundance. Uranus and Neptune formed even farther out, in regions where cold temperatures allowed thick layers of ice and gas to gather around smaller rocky cores. Distance from the Sun determined not only temperature but also the density of available matter, with the outer disk containing less material per unit volume than the inner regions.

Inside the disk, particles collided constantly in a dance of construction and destruction. At first they were tiny grains, no larger than dust motes, measuring mere micrometers across. These grains moved through the disk on slightly different orbits, sometimes brushing against each other at gentle speeds. When they collided softly enough, electrostatic forces and van der Waals interactions caused some grains to stick together. This process, called accretion, built larger particles grain by grain. Over thousands of years, these grains became pebbles. Over tens of thousands of years, pebbles became rocks. Over hundreds of thousands of years, rocks became boulders measuring meters across. As these bodies grew larger, a new force entered the process. Gravity, negligible for dust grains, became significant for objects the size of mountains. Bodies measuring kilometers in diameter exerted enough gravitational pull to actively attract more material, pulling in smaller objects that passed nearby. The largest of these growing worlds became planetesimals, solid bodies ranging from kilometers to hundreds of kilometers across. They were the first substantial building blocks of planets, massive enough to carve out zones in the disk and dominate their neighborhoods. Over millions of years, planetesimals collided violently, their impacts generating tremendous heat. Some collisions shattered both bodies, but many resulted in mergers that created even larger objects. This chaotic process gradually concentrated matter into fewer and fewer bodies, each one growing more massive through a combination of gravitational attraction and direct collision.

Earth began as one of these growing bodies, forming in the inner solar system where rocky materials dominated. It assembled through repeated collisions with other planetesimals and smaller objects over tens of millions of years. Each impact added mass and heat, the kinetic energy of collision converting into thermal energy that melted rock and metal. The growing Earth was a molten world, its surface repeatedly liquefied by the violence of constant bombardment. During this period, dense materials like iron and nickel sank toward the center under gravity, while lighter silicates floated toward the surface, a process called differentiation that gave Earth its layered structure. One collision in particular was so massive that it reshaped the forming Earth entirely. A Mars-sized object, sometimes called Theia by astronomers, struck the young planet in a glancing blow. The impact occurred perhaps fifty to one hundred million years after the solar system began forming. The collision was catastrophic beyond imagination, releasing energy equivalent to billions of nuclear weapons. It melted much of Earth's surface and sent an enormous spray of vaporized and molten material into orbit around the planet. This debris, a mixture of material from both Earth and the impactor, formed a ring around Earth. Through the same process of accretion that built the planets, this orbiting debris gathered together over thousands of years and eventually became the Moon. The Earth that emerged from this collision was hotter, larger, and already accompanied by the satellite that would later stabilize its rotation and moderate its axial tilt over geological time.

While the inner planets took shape from rock and metal, the giant planets developed through a fundamentally different process. Jupiter formed in the region beyond the frost line where ices swelled the amount of available solid material. It began like Earth, as a growing core of rock and ice perhaps ten to fifteen times Earth's mass. But once it reached this critical size, its gravity became strong enough to capture gas directly from the surrounding nebula. Jupiter grew rapidly, gathering not only solid material but also vast quantities of hydrogen and helium, the lightest and most abundant elements in the disk. Its enormous gravity allowed it to sweep up gas faster than the disk could clear, and it became the largest planet in the solar system, with a mass more than three hundred times that of Earth. Saturn followed a similar path, forming slightly farther from the Sun. It too built a solid core that captured hydrogen and helium, though it gathered less gas than Jupiter, perhaps because it formed slightly later or because less material was available at its greater distance. Farther out still, Uranus and Neptune formed from smaller cores that captured thick layers of ice and gas. These ice giants grew more slowly than Jupiter and Saturn. Their distance from the Sun placed them in regions where the disk was less dense and orbital periods were longer, meaning fewer collisions occurred per unit time. They captured some hydrogen and helium but retained much larger proportions of heavier ices, giving them different compositions from their larger cousins. Their final size and composition reflect the limited material available in the outer reaches of the solar nebula.

As the planets grew toward their final masses, the remaining gas in the disk began to thin and disappear. The young Sun entered an active phase in its early life, emitting intense radiation and powerful stellar winds. These winds consisted of charged particles streaming outward from the Sun's surface at hundreds of kilometers per second. The radiation and particle streams exerted pressure on the gas in the disk, particularly the lighter hydrogen and helium. Over just a few million years, this solar wind blew much of the remaining gas away from the inner solar system, sweeping it out into interstellar space. Winds of charged particles cleared the space around the planets and halted further growth by removing the raw material from which planets form.

Once the gas dispersed, planetary growth essentially ended except through direct collision. The leftover solid material continued to orbit as smaller bodies, unable to accumulate into planets either because they formed too slowly or because gravitational perturbations prevented them from merging. Asteroids remained in the region between Mars and Jupiter, a population of rocky bodies ranging from dust particles to objects hundreds of kilometers across. They were unable to form a planet because Jupiter's gravity disrupted their attempts to merge, stirring their orbits and causing collisions to occur at speeds too high for accretion. Instead of building a planet, the asteroid belt remained a collection of fragments, a museum preserving samples of the early solar system. Icy bodies settled in distant regions beyond Neptune, forming the Kuiper Belt and the more distant Oort Cloud, vast reservoirs of comets that still orbit the Sun in elliptical and highly inclined paths. These frozen remnants contain pristine material from the solar nebula, largely unchanged since the solar system's formation.

Earth formed in a region too warm for ice to survive, yet it still ended up with abundant water. This happened because icy objects from beyond the frost line were repeatedly knocked inward by the early movements of Jupiter and Saturn. Many of these bodies collided with the young Earth over millions of years, delivering water locked inside their mineral structures and frozen interiors. Comets and water-rich asteroids acted as couriers, gradually supplying the material that later became Earth's oceans.

The final shape of the solar system emerged gradually over hundreds of millions of years after the planets formed. Initially, the orbits were more chaotic, with planets migrating inward or outward through gravitational interactions with the disk and with each other. Jupiter may have moved inward and then outward again, shepherding material and sculpting the distribution of smaller bodies. Neptune likely formed closer to the Sun than its current position and migrated outward, scattering icy planetesimals as it moved. These migrations had profound effects on the smaller bodies, triggering bombardment events and shaping the current architecture. Eventually, the planets settled into stable orbits, each one occupying a zone where its gravitational influence dominated and where the combined pull of the Sun and other planets created long-term stability. The Sun itself stabilized as a mature star, its core fusion reaching equilibrium with gravity. It began shining with the steady light that has illuminated the solar system for more than four billion years. The solar system became a collection of worlds, each with unique characteristics shaped by its distance from the Sun, its size, its history of collisions, and the materials available during its formation. Rocky worlds orbited close to their star. Gas and ice giants circled farther out. Smaller bodies filled the spaces between and beyond, and a single star held the entire system in its gravitational embrace.

The solar system's formation was not a peaceful event unfolding in quiet isolation. It was a period of intense heat, violent impacts, and constant change stretching across tens of millions of years. Temperatures ranged from thousands of degrees near the young Sun to near absolute zero in the outer reaches. Collisions shattered worlds and built new ones from the fragments. Yet from this chaos emerged a structure that supports life on at least one world and possibly others. The conditions required for Earth to form as a habitable planet were shaped by processes that stretched across billions of years and depended on countless factors aligning. The Sun had to form with the right mass, not so large that it would burn out quickly and not so small that its habitable zone would be too close and its lifetime too long. The disk had to cool at the right rate and contain the right mixture of elements. The Earth had to form at the right distance from the Sun, in a zone where temperatures allowed liquid water to exist on its surface. It had to gather enough volatile materials like water and carbon while also accumulating sufficient mass to hold an atmosphere through gravity.

The Moon had to form through the giant impact, creating a satellite large enough to stabilize Earth's axial tilt and prevent dramatic climate swings. Each step was part of a long chain of events that began with a collapsing cloud of gas, each one necessary but not sufficient alone for the eventual emergence of life.

Understanding the formation of the solar system reveals that Earth is not isolated or accidental, nor is it necessarily unique. It is part of a natural process that occurs throughout the galaxy wherever clouds of gas and dust exist. Stars form with disks, and planets grow within those disks following the same physical laws that governed our own system's formation. Observations of other young stars confirm this, showing protoplanetary disks in various stages of planet formation. Some systems will resemble ours, with rocky worlds near the star and gas giants farther out. Others will differ dramatically, with giant planets orbiting close to their stars or systems containing planets unlike any in our solar system. The formation of the solar system is one chapter in the larger story of how the universe creates environments where matter can organize itself into worlds. It demonstrates that planetary systems are common outcomes of star formation, and it provides context for understanding our place in the cosmos.

With the solar system formed and stabilized, the stage was set for the rise of geological activity, atmospheric development, and eventually life. The next chapters explore how Earth developed its distinctive features, how plate tectonics and volcanism shaped its surface, how an atmosphere capable of supporting life emerged, how life itself began in Earth's oceans, and how the universe shaped everything that followed. The process that began with a collapsing cloud of gas led ultimately to a world capable of supporting oceans, continents, living creatures, and human beings capable of looking back and understanding their own origins. The solar system's formation was the necessary first step in a chain of events that produced not only our world but also the minds capable of comprehending how it all began.

Chapter Eighteen
Our Planet and the Cosmos

The Earth is a small world in a vast universe, yet everything about it depends on events that happened long before the planet formed. To understand Earth and the other planets, we must look back to the moment when the Sun was young, roughly four and a half billion years ago. Before any planets existed, the Sun was surrounded by a wide disk of gas, dust, and rocky grains left over from the cloud that collapsed to form it. This disk, known as the protoplanetary disk or solar nebula, contained all the raw materials that would eventually become the planets, moons, asteroids, and comets. The disk slowly flattened as it spun, responding to the fundamental forces of rotation and gravity. Inside it, countless particles moved in orbits around the young star, colliding constantly at speeds that sometimes destroyed them and sometimes caused them to stick together. These particles formed larger clumps through a process called accretion, and those clumps grew into the first building blocks of planets, known as planetesimals. The story of Earth begins inside this swirling disk of material that circled the newborn Sun, where physics and chemistry worked together to transform dust into worlds.

As the disk cooled over the first few million years, materials separated based on their ability to withstand heat. This separation followed what scientists call the condensation sequence, a pattern determined by the melting and freezing points of different substances. Close to the Sun, where temperatures exceeded a thousand degrees, only metals like iron and nickel and rocky substances like silicates could survive in solid form. Everything else remained as gas or was blown away by the intense radiation and solar wind from the young star. Farther away, beyond a boundary known as the frost line or snow line, temperatures dropped low enough that ice and lighter elements could remain solid.

Not every star creates a habitable zone like the one Earth occupies. Smaller stars called red dwarfs have habitable zones so close to the star that planets there often become tidally locked, with one side in permanent daylight and the other in endless night. Larger stars, meanwhile, burn their fuel so rapidly that their habitable zones move outward over time, giving planets little stability to develop long-lasting environments. The Sun's size places it in a rare middle category—large enough to provide a wide, stable region where liquid water can exist, yet small enough to shine for billions of years. Earth's position within this slowly shifting zone is one of the reasons its surface stayed warm enough for oceans to form and remain liquid over geological time.

Water ice, ammonia ice, and frozen methane survived in these cooler regions, vastly increasing the amount of solid material available to build planets. This simple difference in temperature, determined purely by distance from the Sun, shaped the fundamental character of the planets. The inner solar system formed rocky worlds like Mercury, Venus, Earth, and Mars, sometimes called the terrestrial planets. These bodies are relatively small and dense, composed primarily of metal and rock. Farther out, where temperatures were lower and volatile materials could condense, giant planets formed from gas and ice. Jupiter and Saturn became gas giants, accumulating vast envelopes of hydrogen and helium around rocky cores. Uranus and Neptune, forming even farther out where the disk was less dense, became ice giants with smaller gas envelopes but significant amounts of water, ammonia, and methane. Gravity pulled material together wherever mass concentrated, and with each collision, the growing bodies became more massive, their increasing gravity attracting even more material. Over millions of years, these collisions and accumulations created the diverse family of planets we know today.

Earth formed through these repeated impacts, growing larger with each addition. Countless rocks and smaller bodies, some no larger than pebbles and others the size of mountains, merged to create a growing sphere. The process was violent and chaotic. As planetesimals collided, their kinetic energy converted to heat, and the young Earth grew steadily warmer. The planet differentiated during this time, with heavier elements like iron sinking toward the center to form the core while lighter silicate rocks floated upward to form the mantle and eventually the crust. One enormous collision with a Mars sized object, sometimes called Theia, melted the young Earth entirely and sent an enormous spray of debris into orbit. That debris, made of material from both Earth and the impactor, eventually gathered through its own gravity to form the Moon. The impact was so energetic that it liquefied the outer layers of both bodies, and the resulting debris disk glowed white hot as it orbited the molten Earth. The Earth that emerged from this early chaos was hot, unstable, and covered with molten rock. Its surface was a magma ocean that slowly solidified as heat radiated into space. Gradually, over tens of millions of years, it cooled enough to form a solid crust. Volcanoes punctured this crust continually, releasing gases that had been trapped in the interior. These volcanic emissions formed the first atmosphere, rich in water vapor, carbon dioxide, nitrogen, and other gases but lacking the free oxygen that would come much later. Water, delivered by countless comets from the outer solar system and released from minerals within the planet itself, collected in low areas as the surface cooled below the boiling point. The first oceans formed, covering much of the young planet. The conditions that allowed life to appear were already taking shape in these warm, mineral rich waters beneath a thick, steamy atmosphere.

Planets across the universe form in similar ways, following the same physical principles that governed the formation of our solar system. Wherever a young star has a disk of material around it, a natural consequence of the way stars form from collapsing clouds, planets begin to grow through accretion. Astronomers have detected thousands of these protoplanetary disks around stars in our galaxy, as well as thousands of exoplanets orbiting distant suns. Some planets remain small and rocky, similar to Earth and Mars. Others gather massive envelopes of gas and become giants that dwarf Jupiter. Some form moons through violent collisions, while others capture moons later from passing bodies or accumulate them from smaller debris disks. The laws of physics guide every step regardless of where it happens. Gravity pulls material inward, causing matter to clump. Rotation causes disks to flatten into thin planes where collisions are more frequent. Repeated collisions build planets from small beginnings through a gradual accumulation of mass. Heat from these impacts and from radioactive decay shapes the internal structure of growing worlds. These same processes occur in star systems throughout the galaxy and indeed throughout the universe, making planets a natural and probably inevitable part of cosmic evolution. The formation of worlds is not a rare accident but a common outcome wherever the right conditions exist.

The tides created by the Moon's gravitational pull did more than move water. They repeatedly exposed and submerged coastal environments, creating natural cycles of wetting and drying that concentrated organic molecules. Many scientists believe these tidal zones acted as chemical laboratories where early life experimented with new structures. Pools that filled and evaporated over and over allowed simple molecules to link into more complex chains, providing one possible pathway toward the first self-replicating systems. Without the Moon's strong tidal influence, Earth's early coastlines would have been far less dynamic, and the chemistry of life may have taken a very different path.

Earth is unique to us because it is our home, but the conditions that make it special are the result of a series of natural events, not divine intervention or extraordinary cosmic luck. Its distance from the Sun, roughly 150 million kilometers, places it in what astronomers call the habitable zone, where temperatures allow liquid water to exist on the surface. Water is essential for life as we know it, serving as a solvent for chemistry and a medium for biological processes. Earth's magnetic field, generated by the motion of liquid iron in its outer core, protects the planet from harmful radiation carried by the solar wind. Without this shield, the atmosphere would slowly erode away into space, as happened on Mars when its magnetic field weakened. Earth's atmosphere holds enough heat through the greenhouse effect to keep temperatures stable, preventing the planet from freezing despite receiving only a fraction of the Sun's total energy output. The atmosphere also provides pressure that keeps water liquid across a wide range of temperatures. Earth's rotation, completing one turn roughly every twenty four hours, creates the cycle of day and night, preventing one side from overheating while the other freezes. Its axial tilt of about twenty three degrees creates seasons as different hemispheres lean toward or away from the Sun during the annual orbit. The presence of the Moon stabilizes the planet's tilt over geological time, preventing wild swings in axial orientation that would cause catastrophic climate changes. The Moon also creates tides through its gravitational pull, stirring the oceans and possibly playing a role in the development of early life in tidal zones. These features appeared through the ordinary processes of planetary formation and orbital dynamics, shaped by gravity, thermodynamics, and the composition of the solar nebula, yet together they created a world capable of supporting complex organisms. Earth is not the only planet in the universe with such conditions, but it is the one where those conditions led to the remarkable diversity of life we see today.

Beyond the planets, the solar system sits inside a much larger structure, one that defines our place in the cosmos. Our Sun is one of hundreds of billions of stars in the Milky Way galaxy, a vast collection of matter held together by gravity. The galaxy is a vast spiral of stars, gas, and dark matter that rotates slowly around its center, where a supermassive black hole millions of times more massive than the Sun anchors the structure. The Milky Way spans about one hundred thousand light years from edge to edge, though most of its mass is concentrated in the bright central bulge and the disk that surrounds it. Stars move along curved paths shaped by the combined gravitational pull of all the mass in the galaxy. The Sun orbits the galactic center at a speed of roughly 200 kilometers per second, yet it takes about 225 million years to complete one orbit, a period sometimes called a galactic year. Some regions of the galaxy are crowded with stars, particularly in the dense clusters near the center and in the globular clusters that orbit in the halo. Others are mostly empty, especially the spaces between the spiral arms. Between the stars lies gas, primarily hydrogen, that will one day collapse under its own gravity to form new star systems. Dust grains and molecules drift through this interstellar medium, creating dark clouds where stars are born. The Milky Way is not static. It evolves continuously as stars are born in the spiral arms, live out their lives over millions or billions of years, and eventually die, returning material enriched with heavier elements to the galaxy to be used again in the next generation of stars and planets. This cycle of birth, life, death, and rebirth has been ongoing for most of the universe's history.

The Milky Way is only one galaxy among many, a single participant in a much larger cosmic community. The universe contains hundreds of billions of galaxies, each with billions or trillions of stars, spread across observable space that extends more than forty billion light years in every direction. Some galaxies are spirals like ours, with rotating disks and curved arms where new stars form. Others are elliptical, appearing as smooth, featureless balls or ellipses of stars with little gas and almost no ongoing star formation. Still others are irregular, lacking any clear structure, often shaped by past collisions or gravitational encounters. They are spread across the cosmos in patterns influenced by gravity and the invisible scaffolding of dark matter, which provides most of the gravitational pull in the universe. Galaxies are not distributed randomly but cluster together in groups and larger structures. The Milky Way belongs to a small collection called the Local Group, which includes the Andromeda Galaxy, the Triangulum Galaxy, and dozens of smaller satellite galaxies. The space between galaxies is immense, typically millions of light years, and yet they all move within the expanding fabric of the universe, carried along by the stretching of space itself. Over time, galaxies collide and merge, drawn together by gravity despite the expansion. When they do, their stars rarely collide because of the vast distances between them, but their gas clouds combine, triggering bursts of new star formation. Their structures change dramatically through these interactions. The universe is always shifting and evolving, even on scales far beyond human experience or comprehension.

The large scale structure of the universe forms a web, sometimes called the cosmic web, that reveals the fundamental architecture of space.

The Milky Way also has a distant but dramatic future. It is slowly moving toward the Andromeda Galaxy, another massive spiral in the Local Group. Although they are separated by more than two million light years, gravity is pulling them together, and the two galaxies will merge in roughly four billion years. When this happens, their stars will swirl into new patterns, their gas clouds will collide and ignite bursts of star formation, and the combined system will eventually settle into a single, larger galaxy. Despite the scale of this event, individual stars—including our Sun—are unlikely to collide, because the space between stars is vast. The merger is a reminder that galaxies are dynamic, ever-changing structures shaped by gravity even across enormous distances.

Galaxies cluster along strands and filaments of this web, separated by enormous regions of emptiness called voids where very few galaxies exist. These voids can be hundreds of millions of light years across, representing the largest empty spaces in the universe. The filaments, by contrast, are dense rivers of galaxies flowing through space, connecting massive clusters where thousands of galaxies congregate. These patterns reflect the tiny variations in density that were present in the earliest moments after the universe began, imprinted on the fabric of space during the first fraction of a second. Gravity amplified those variations over billions of years, drawing matter together wherever density was slightly higher and emptying regions where it was slightly lower. The web continues to evolve, shaped by the flow of dark matter, which provides the gravitational framework, and by the expansion driven by dark energy, which pushes everything apart. Although this structure appears frozen when viewed over a human lifetime or even over recorded history, it changes slowly but continuously over billions of years as galaxies flow along the filaments, clusters grow larger, and voids expand.

Earth, the solar system, the Milky Way, and the universe all share a common story of formation and evolution. They formed from the same basic ingredients, matter and energy released from the Big Bang, shaped by the same laws of physics that govern gravity, thermodynamics, and quantum mechanics, and guided by the same unfolding of time and space that has been in progress for nearly fourteen billion years. Our planet is part of a larger structure that stretches far beyond what we can see with our eyes or even with our most powerful telescopes. Yet everything that exists on Earth, from the deepest oceans to the highest mountains to every living creature that has ever existed, depends on the formation of stars and galaxies that occurred long before our world emerged from the solar nebula. The atoms in our bodies were forged in the cores of ancient stars that exploded before the Sun existed. The water in our oceans came from ice that condensed in the cold outer reaches of the solar system. The iron in our blood sank to Earth's core during the planet's violent youth. We are, quite literally, made of stardust, shaped by cosmic processes that began at the universe's birth.

Understanding Earth in this cosmic context reveals something important about our place in nature. The planet is not isolated or separate from the rest of the cosmos. It is one chapter in a story that began with the first particles emerging from the Big Bang, continued through the formation of atoms in the first minutes, progressed through the creation of stars and galaxies over billions of years, and expanded into the vast structures that fill the observable universe. Every stage built upon the previous one, creating the conditions for the next. The stars created the elements. The elements formed planets. The planets provided the environments where, on at least one world we know of, chemistry became biology and life began its long journey toward complexity. The next step in this journey is to explore how we know these things, how we pieced together this grand narrative from observations and experiments. The universe leaves clues in light, matter, and the structure of space itself, and by studying these clues with careful attention and scientific rigor, we uncover the story of our origins and our connection to everything that exists.

Chapter Nineteen
How We Know What We Know

Understanding the universe might seem impossible at first glance. The events that shaped reality happened billions of years ago, long before Earth existed, and far beyond anything we can reach with our hands. The stars we see at night exist at distances so vast that their light takes years, centuries, or even millennia to reach our eyes. The atoms that compose our bodies were forged in stellar furnaces that burned out before our sun was born. Yet despite these challenges, science has uncovered a detailed history of the cosmos, from the birth of the first particles in the intense heat of the early universe to the slow assembly of galaxies across billions of years. It has revealed the age of the Earth with precision, mapped the internal structure of atoms down to their smallest components, explained the nuclear reactions that power stars, and traced the gradual emergence of life from simple chemistry to the complexity we see today. The way we know these things is not through guesswork or imagination, nor through ancient stories passed down through generations. We know them because the universe leaves evidence behind, and that evidence can be read with remarkable accuracy. Every process in nature leaves its mark. Every event writes its signature into the fabric of reality. The story of the universe is written in light, matter, and time, and scientists have learned how to interpret that story with increasing sophistication.

One of the most important tools in understanding the universe is light itself. Light is not merely illumination. It is a carrier of information across vast distances, a messenger that travels through the emptiness of space at three hundred thousand kilometers per second. When a star shines, it sends out light that travels through space until it reaches us, sometimes journeying for millions or billions of years before entering our telescopes. Light spreads outward in straight paths unless something bends it through gravity or refracts it through matter, and it carries within itself the fingerprints of the object that produced it. When light passes through a prism or enters a scientific instrument called a spectrograph, it separates into its component colors, spreading out into a spectrum much like a rainbow. This spectrum is not simply beautiful. It is informative. Each chemical element absorbs and emits light at specific wavelengths, creating unique patterns of bright and dark lines in the spectrum. Hydrogen creates one pattern. Helium creates another. Iron, carbon, oxygen, and every other element each produce their own distinctive signature. By studying these patterns in the light from distant stars and galaxies, scientists can determine precisely what those objects are made of, even though they exist trillions of kilometers away. The same spectral analysis reveals the temperature of a star by examining which wavelengths it emits most strongly. Hot stars shine with a blue white light, while cooler stars glow with a reddish hue. The patterns also reveal how fast an object is moving through space, and when combined with other observations, they help us estimate how far away that object might be. This method, called spectroscopy, allows us to study objects we could never visit and gives us a clear window into both the composition and the behavior of the distant cosmos.

Light does more than show us what distant objects are made of. It reveals how the universe changes and evolves. When a galaxy is moving away from us, the light it emits becomes stretched by the expansion of space itself. This stretching shifts the light toward longer wavelengths, moving the entire spectrum toward the red end. Scientists call this phenomenon redshift. The amount of shift reveals the speed at which the galaxy is receding from us. When astronomers in the early twentieth century began measuring the light from distant galaxies, they discovered something unexpected. Most galaxies showed this redshift, which told them that the universe is expanding. The farther a galaxy is from us, the faster it appears to move away. This relationship between distance and velocity is not random. It follows a predictable pattern that tells us something profound about the nature of space itself. Space is not static. It is growing, stretching, carrying galaxies along with it like dots on the surface of an inflating balloon. This simple observation, first recognized clearly by Edwin Hubble, supports the idea that the universe had a beginning, a moment when all of space was compressed together. The same light also shows us how galaxies evolve and change over time. By observing galaxies at different distances, we are actually seeing them at different stages of their development. This works because light travels at a finite speed. When we observe a galaxy ten billion light years away, we are seeing light that left that galaxy ten billion years ago. We are literally looking backward through time. Distant galaxies appear younger, less organized, often more active and chaotic than nearby galaxies. In this way, looking far into space allows us to see far into the past, giving us a direct view of cosmic history unfolding across billions of years.

The age of the universe itself is measured through several independent pieces of evidence, each one confirming the others. One of the strongest comes from what scientists call the cosmic microwave background. This faint glow fills the sky evenly in every direction, detectable not with visible light but with radio telescopes sensitive to microwave radiation. It is the cooled radiation left over from the early universe, a remnant of a time when the cosmos was so hot and dense that light couldn't travel freely through it. As the universe expanded and cooled, this radiation was released, and it has been traveling through space ever since, growing cooler and more stretched with each passing billion years. Today it has a temperature of just 2.7 degrees above absolute zero, cold and faint but still detectable. By measuring its temperature and studying the tiny variations within it, variations that represent small differences in density from one region to another in the early universe, scientists can calculate how long the universe has been expanding. They can trace the growth of space backward through time until they reach the point where everything we see today was compressed into an unimaginably hot and dense state. These calculations reveal that the universe is about thirteen point eight billion years old, with an uncertainty of only a few tens of millions of years. This number is not guessed or estimated loosely. It comes from precise measurements that agree with one another across different methods, all pointing to the same conclusion. The expansion rate measured from distant galaxies, the temperature of the background radiation, and the abundances of light elements all fit together into a consistent picture of cosmic history.

The age of the Earth is known through a different but equally reliable method. Certain elements found inside rocks are naturally unstable. Their atomic nuclei contain an imbalance of protons and neutrons that makes them prone to decay, transforming gradually into other, more stable elements. This process happens at steady, unchanging rates. A uranium atom will decay into lead over a span of millions or billions of years, ticking away like an atomic clock. Importantly, these decay rates don't vary with heat, pressure, chemical environment, or any other surrounding conditions. They are governed entirely by the internal structure of the nucleus itself, making them extraordinarily reliable timekeepers. By measuring how much of a radioactive element remains in an ancient rock and how much has transformed into its decay product, scientists can determine how long the rock has existed. Rocks from the oldest surfaces on Earth, found in places like Western Australia and Greenland, yield ages of more than four billion years. Meteorites that formed at the same time as the planets, before they fell to Earth, show similar ages clustered around four and a half billion years. Moon rocks brought back by the Apollo missions also fit this timeframe. Together, these clues reveal when the solar system took shape, marking the moment when gravity pulled together dust and gas into the sun, planets, moons, and asteroids.

Understanding atoms and the particles that compose them comes from experiments that explore the smallest scales accessible to science. Scientists use machines called particle accelerators that propel subatomic particles to enormous speeds, close to the speed of light itself. These particles are then directed to collide with one another or with stationary targets. When particles collide at such high energies, they break apart briefly or transform into other particles, revealing the components and forces that make them up. Surrounding the collision point are sophisticated detectors that measure the paths, energies, and identities of the resulting fragments. Some particles live for only a tiny fraction of a second before decaying, but their presence can still be inferred from the trails they leave behind. By studying thousands or millions of collisions and analyzing the patterns that emerge, scientists discover the fundamental rules that govern particle behavior and the forces that act on them. These experiments have revealed the existence of quarks, the building blocks of protons and neutrons. They have discovered neutrinos, ghostly particles that pass through matter almost without interacting. They have confirmed the existence of force carrying particles like photons, which transmit the electromagnetic force, and gluons, which hold quarks together inside protons and neutrons. These findings have shown us how particles interact with one another and how matter holds itself together at the most basic level. The same experimental approach has allowed scientists to test theories, predict new particles, and confirm those predictions through careful measurement.

Evidence for evolution and the history of life on Earth comes from multiple converging lines of investigation, including fossils, genetics, anatomy, and the layers of rock beneath our feet. Fossils are the preserved remains or impressions of ancient organisms, and they show directly how life forms changed over vast stretches of time. Older rock layers, deposited hundreds of millions or billions of years ago, contain simpler life forms. Single celled organisms appear first. Later, more complex organisms emerge in the fossil record. Fish appear before amphibians. Amphibians appear before reptiles. This layered sequence is not random. It represents a progression through time. Genetics adds another dimension to this story. DNA, the molecule that carries hereditary information, changes slowly over generations through mutations. By comparing the DNA sequences of different organisms, scientists can measure how closely related they are. Humans and chimpanzees share nearly ninety nine percent of their DNA, indicating a recent common ancestor. Humans and mice share less, and humans and fish share even less still. The pattern matches what fossils predict. The anatomical similarities between species also tell a story. The bones in a human arm, a whale's flipper, a bat's wing, and a horse's front leg all follow the same basic structure, modified for different purposes. This shared architecture reveals common ancestry. These methods, when taken together, show that all life on Earth shares a common origin, arising from the same primordial chemistry billions of years ago. They demonstrate that the complexity and diversity of life increased gradually over time, shaped by natural selection, genetic variation, and environmental changes across countless generations.

The methods of science work because they rely on observations that can be tested, repeated, and verified by independent investigators. Measurements don't depend on belief, opinion, authority, or cultural tradition. They depend on evidence that exists whether anyone believes in it or not. A spectral line appears at a certain wavelength because of the physics of atoms, not because someone wants it to. A radioactive element decays at a fixed rate regardless of who measures it. When different methods point to the same result, confidence in that result grows stronger. When the age of the universe calculated from the expansion rate matches the age calculated from the cosmic background radiation, we gain assurance that both methods are revealing something true about reality. When new evidence appears that contradicts an existing idea, scientific theories are refined, adjusted, or sometimes replaced, but the underlying method remains the same. Science is not a collection of fixed answers handed down without question. It is a process of asking questions, gathering evidence, testing ideas against observation, and refining our understanding based on what the universe shows us. This process has allowed us to uncover a story far greater, more detailed, and more astonishing than anyone could have imagined without these tools. It has revealed a universe billions of years old, filled with trillions of galaxies, governed by comprehensible laws, and capable of producing stars, planets, and living organisms through natural processes.

The universe may seem distant, abstract, and unreachable, existing in scales of time and space beyond ordinary human experience. Yet everything in it leaves traces that we can detect and interpret. Light carries information across the void of space, bringing us news from the edges of the observable universe. Rocks preserve the chemical and physical history of Earth, locking the record of billions of years into their crystalline structures. Fossils record the evolution of life, showing us creatures that walked, swam, or flew millions of years before humans existed. Particles reveal their secrets through carefully designed experiments, allowing us to probe the fundamental nature of matter and energy. The universe speaks through these clues, and science is the method, the language, we use to listen. By following this evidence carefully, checking our interpretations against multiple lines of inquiry, and remaining open to correction when new data arrives, we can understand how everything came to be. We can see our place in the larger structure of existence, not as the center of creation but as part of an immense and ancient cosmos that operates according to discoverable principles.

As we move on, we will shift our focus from how we know about the universe to what the universe is doing now and where it may be heading. The future of the cosmos is shaped by the same physical laws that shaped its beginning, the same forces that ignited the first stars and assembled the first galaxies. These laws reveal a future filled with changes as remarkable and profound as anything that has occurred in the past, a future that extends far beyond the lifespan of our sun, our galaxy, and perhaps even matter itself.

Chapter Twenty

The Universe's Future

The universe we live in is vast, ancient, and always changing. When we look at the night sky, we see only a tiny part of what exists. The nearest star beyond our Sun lies more than four light years away, meaning its light has traveled through space for over four years before reaching our eyes. The farthest galaxies we can observe lie billions of light years distant, so far that we see them as they appeared when the universe was still young. Stars appear as small points of light, yet each one is a sun in its own right. Some are larger than our Sun and some are smaller. Some are young and bright, burning their fuel rapidly and shining with fierce intensity. Others are nearing the end of their lives, having spent billions of years converting hydrogen into heavier elements. Between the stars lies the darkness of space, but even that darkness is filled with invisible matter, energy, and distant galaxies too faint for our eyes to detect without powerful telescopes. The universe today is the result of billions of years of expansion, evolution, and transformation. To understand its future, we must first understand how it behaves now.

The universe continues to expand, just as it has since the moment of its beginning. Galaxies move away from one another, carried by the stretching of space itself. This expansion is not caused by galaxies speeding through space like ships sailing through an ocean. Instead, space itself grows, increasing the distances between galaxies much as the surface of an inflating balloon grows and carries dots drawn on it farther apart. Every point in space moves away from every other point. An observer in any galaxy would see all other distant galaxies receding, creating the illusion that they occupy a special central position. In truth, there is no center to this expansion. It happens everywhere at once.

Observations made over recent decades show that this expansion is not slowing down as scientists once expected. It is speeding up. Distant galaxies are accelerating away from us at an ever increasing rate. Something in the universe pushes space outward with a gentle but persistent force. Scientists call this mysterious influence dark energy. Although we don't yet know what dark energy is or what causes it, measurements suggest it makes up nearly seventy percent of all the energy in the universe. Its presence affects the future shape and size of the cosmos in profound ways. Dark energy appears to act as a kind of antigravity, not pulling matter together but pushing space apart. Unlike gravity, which weakens with distance, dark energy seems to have the same strength everywhere, filling the fabric of space uniformly.

Galaxies gather in clusters and superclusters, connected by long strands of dark matter that stretch across hundreds of millions of light years. These structures form a cosmic web that spans the entire observable universe. If we could see the universe from far outside, it would resemble a vast sponge or foam, with galaxies concentrated along filaments and sheets while enormous voids lie between them. Gravity binds galaxies together inside these clusters, holding them in place despite the expansion of space. Within a galaxy cluster, gravity is strong enough to resist the pull of dark energy. But between the clusters lies enormous emptiness, regions where gravity is too weak to counteract expansion. As the universe expands faster, these distances grow larger. Over time, galaxies that are not bound by gravity will drift farther apart until they disappear beyond one another's cosmic horizons. The night sky in distant eras, trillions of years from now, will look very different from the sky we see today. Observers in that far future would see only their own galaxy and perhaps a few neighbors, with the rest of the universe having receded beyond reach.

Stars continue to be born in clouds of gas and dust scattered throughout galaxies. These clouds, sometimes spanning hundreds of light years, collapse under their own gravity when disturbed by shockwaves from nearby supernova explosions or by collisions with other clouds. As a cloud contracts, it fragments into smaller pieces. Each fragment grows denser and hotter until nuclear fusion ignites in its core, and a new star is born. This process often creates planetary systems as well, with leftover material forming disks that gradually coalesce into planets, moons, and asteroids. The process of star formation will continue for many billions of years, but not forever. Eventually, the gas needed to create new stars will be used up or locked away in stellar remnants that no longer return material to the cosmos. Galaxies will become quieter as fewer young stars appear to replace those that die. The brilliant blue stars that currently mark regions of active star formation will vanish. Older, cooler stars will dominate for a time, giving galaxies a reddish hue. Then even these will fade and die, leaving behind white dwarfs, neutron stars, and black holes. These remnants will linger for unimaginable lengths of time, slowly cooling as they radiate away their remaining heat into the expanding void.

The fate of the universe depends on how expansion continues and on the nature of dark energy. If dark energy remains constant in strength, as current observations suggest, the universe will grow colder and emptier as time passes. Galaxies outside our local group will move beyond one another's reach, their light stretched to wavelengths so long that they become undetectable. Stars will burn out one by one over timescales of trillions of years. Small red dwarf stars, the most numerous and longest lived, will be the last to fade. When they finally exhaust their fuel, the universe will enter what scientists call the degenerate era, a time when only stellar remnants remain. Over extremely long periods measured in numbers almost beyond comprehension, matter itself may break down into simpler forms. Protons, once thought to be eternal, might slowly decay into lighter particles. The universe would then drift into a state of quiet darkness, a slow fading of everything that once shined. This scenario, sometimes called heat death, represents a universe that reaches maximum entropy, where all useful energy has been dispersed and no further change is possible.

Another possibility is that dark energy could change over time. If its strength increases, as some theoretical models suggest, expansion would speed up even more dramatically. Space would stretch faster and faster, eventually pulling apart not just galaxy clusters but individual galaxies, then solar systems, then planets, and finally even atoms themselves. This scenario, known as the Big Rip, would occur if dark energy grows strong enough to overcome all forces, even the nuclear forces that hold atoms together. The universe would be torn apart in a final moment of violent expansion. This outcome is far from certain, and current evidence does not strongly support it, but it remains one of the possibilities that scientists must consider when contemplating the ultimate fate of the cosmos.

There is also the question of what will happen to black holes, those regions of space where gravity is so intense that nothing, not even light, can escape. Over incredibly long stretches of time, black holes slowly lose energy through a process discovered by physicist Stephen Hawking. Quantum effects near the boundary of a black hole allow it to emit faint radiation, causing it to gradually evaporate. For black holes the size of those formed from collapsed stars, this process is almost impossibly slow. A black hole with the mass of our Sun would take far longer than the current age of the universe to evaporate even slightly. Supermassive black holes, millions or billions of times the mass of the Sun, would persist for incomprehensibly longer periods. But if the universe lasts long enough, even the largest black holes will eventually evaporate completely, converting all their mass into radiation. The final stages of this evaporation would be dramatic, with the black hole releasing a burst of energy before vanishing entirely. The end of this process, occurring perhaps 10^100 years in the future, would leave behind a universe filled only with very light particles, photons and neutrinos, drifting through an enormous dark expanse that continues to expand forever.

These possibilities might seem distant and unfamiliar, but they remind us that the universe is not a fixed place frozen in its current form. It has a history that stretches back nearly fourteen billion years, a present that we observe and study, and a future that extends far beyond. The same fundamental laws of physics that shaped the first atoms in the moments after the Big Bang and guided the formation of the first stars continue to shape what comes next. The universe is always in motion, not moving through space but transforming within itself. It grows, changes, and evolves over scales of time far beyond human experience, yet it remains bound by the same principles we can measure and understand in laboratories today.

Despite the vastness of these future eras and the eventual cold fate that may await, the present universe is still vibrant and full of activity. Stars shine brightly across billions of galaxies, fusing hydrogen into helium and releasing the energy that lights up the cosmos. Planets orbit their suns, some bearing atmospheres and oceans, others barren and cratered. Life exists on at least one world that we know of, and possibly on others we have not yet discovered. Galaxies collide and merge in slow motion gravitational dances that span hundreds of millions of years, creating new regions where shock waves compress gas and trigger intense bursts of star formation. Black holes sit quietly in the centers of galaxies, their gravity shaping the orbits of millions of stars around them and occasionally flaring to life when they consume nearby matter. Nebulae glow with the light of young stars, their colors revealing the presence of different elements created in previous generations of stars. Everywhere we look, at every scale from planets to galaxy superclusters, the universe continues to express the same laws of nature that began at the moment of creation.

The future of the universe is written in these same laws, encoded in the properties of dark energy, the behavior of gravity, and the nature of matter and radiation. Whether it becomes a cold and quiet place over timescales we can barely imagine or evolves in ways we don't yet expect, the universe will follow the patterns set by physics. As we continue to study the cosmos with ever more sophisticated instruments and refine our theories with new observations, we gain a clearer understanding of what lies ahead. We cannot change the ultimate fate of the universe, but we can comprehend it, and in that comprehension lies a deeper appreciation for the present moment in cosmic history. The story of the universe is still unfolding across time and space, and we are witnesses to one small but remarkable part of that vast timeline.

In the next chapter, we will explore the future of humanity and the role our species might play in the universe. The cosmos has its own destiny, driven by forces beyond our control, but human beings also make choices that shape their own future and perhaps the future of life itself. Understanding the universe and our place within it helps us recognize both our limitations and our possibilities, grounding us in reality while inspiring us to reach toward the stars.

Chapter Twenty One
Search for Life

For most of human history, people looked at the night sky with curiosity and wondered whether they were alone. The stars appeared distant and unreachable, each one a point of light in a vast darkness. For centuries, these lights seemed fundamentally different from our own Sun, mysterious and perhaps governed by different rules. When astronomers eventually realized that stars were distant suns, each potentially accompanied by worlds of their own, the question of life elsewhere took on new weight. As our understanding of the universe deepened, the question only grew more compelling. Life on Earth arose from a combination of conditions that were shaped by the laws of nature. Those same laws govern the behavior of matter and energy everywhere, from the nearest planets to the most distant galaxies. They determine how atoms combine into molecules, how stars forge heavy elements in their cores, and how chemistry unfolds on planetary surfaces. If Earth could support life under these universal principles, then it is reasonable to ask whether other worlds might have followed similar paths. The search for life beyond Earth is not only a scientific pursuit. It is a continuation of humanity's oldest question about its place in the cosmos.

The search begins with the recognition that life depends on certain basic requirements. On Earth, life needs liquid water, a source of energy, and the right chemical ingredients. Liquid water serves as a solvent that allows complex molecules to move, interact, and form the chemical reactions necessary for biology. It remains liquid across a temperature range that permits intricate chemistry without destroying delicate organic structures. The source of energy can be sunlight, which drives photosynthesis, or chemical energy from reactions between minerals and water. The chemical ingredients include carbon, which forms the backbone of biological molecules, along with hydrogen, nitrogen, oxygen, phosphorus, and sulfur. These elements combine into amino acids, nucleotides, lipids, and sugars that constitute the machinery of life. These conditions allowed the earliest cells to form in the oceans, perhaps near hydrothermal vents where hot water rich in minerals met cooler seawater. They sustained evolution from simple single celled organisms to the remarkable diversity of complex life we see today. When scientists look for life elsewhere, they use these requirements as a guide, not because life elsewhere must be identical to Earth life, but because these are the conditions we understand and know can work. They ask where liquid water might exist, where energy sources might be available, and where the elements needed for life might gather in sufficient concentrations. These questions lead them to examine planets, moons, and even distant worlds orbiting other stars.

Within our own solar system, several places may contain environments suitable for life. Mars stands as the most intensively studied candidate. The planet once had rivers that carved valleys into its surface, lakes that pooled in impact craters, and possibly oceans that covered parts of its northern hemisphere. Evidence appears in dried riverbeds, layered sedimentary rocks that form only in standing water, and minerals such as clays and sulfates that require water to crystallize. Its surface today is dry, cold, and scoured by radiation, but signs of ancient water remain preserved in its rocks and landscape. If life ever existed there during its wetter past, it may have retreated underground as the surface became inhospitable. Liquid water could still be present in aquifers beneath the surface where geothermal heat and pressure from overlying rock maintain temperatures above freezing. Scientists search Mars for chemical signatures such as organic molecules that could indicate past biology, fossil structures that might preserve the shapes of ancient microbes, and minerals like carbonates that form only in the presence of water and that sometimes contain biosignatures. Rovers and landers explore its surface with increasing sophistication, drilling into the soil, analyzing samples with onboard laboratories, and gradually uncovering the planet's history one measurement at a time.

Farther from the Sun, where sunlight dims and temperatures plunge, several icy moons hold oceans beneath their frozen surfaces. These worlds seem unlikely places for life at first glance, covered in ice and orbiting in the cold outer reaches of the solar system. Yet their interiors tell a different story. Jupiter's moon Europa contains a deep ocean of liquid water beneath an ice shell that may be tens of kilometers thick. This ocean exists not because of sunlight but because of tidal forces. As Europa orbits Jupiter, the giant planet's gravity flexes the moon's interior, generating heat through friction in the same way that bending a piece of metal repeatedly makes it warm. This tidal heating keeps the ocean liquid despite the frigid conditions at the surface. Saturn's moon Enceladus reveals its ocean more directly. The moon sprays plumes of water vapor, ice crystals, and organic molecules into space from deep cracks near its south pole. These geysers provide a window into the subsurface ocean without requiring a spacecraft to drill through the ice. By flying through these plumes, probes have sampled the ocean's composition and discovered that it contains salts, organic compounds, and signs of hydrothermal activity. These oceans are kept warm by internal heat rather than sunlight, creating environments that resemble hydrothermal vents on Earth's ocean floors, places where life thrives in complete darkness using chemical energy from the interaction of hot water and rock. If life can exist in such conditions here, sustained by chemistry alone without any need for sunlight, it may exist on these distant moons as well.

Beyond our solar system, the discovery of planets orbiting other stars has expanded the search dramatically. For most of history, the existence of such planets was only a reasonable guess. The first confirmed detection came in the 1990s, and since then, the count has grown to thousands. These exoplanets have been identified through multiple techniques. Some are detected when they pass in front of their stars and cause a slight dimming of starlight. Others reveal themselves through the tiny gravitational wobble they induce in their parent stars. Many of them orbit in regions called habitable zones, distances from their stars where temperatures could allow liquid water to exist on a surface. Some orbit stars similar to the Sun, yellow stars with lifetimes measured in billions of years. Others orbit smaller, cooler red dwarf stars that burn their fuel slowly and may provide stable environments for tens of billions of years, offering far longer timescales for life to develop and evolve. These planets vary widely in size, composition, and atmosphere. Some are gas giants larger than Jupiter. Others are rocky worlds similar to Earth. Still others are super Earths, rocky planets larger than our own with uncertain surface conditions. By studying how starlight passes through their atmospheres during a transit, scientists can identify which wavelengths are absorbed by atmospheric gases. This technique allows them to search for gases such as oxygen, which on Earth is produced by photosynthesis, methane, which can be produced by biological processes, or combinations of gases like oxygen and methane that would react away quickly unless continually replenished by life. These biosignatures represent our best hope for detecting life across interstellar distances.

The search for life is not limited to planets that resemble Earth or organisms that resemble those we know. Some researchers explore the possibility of life based on solvents other than water, such as liquid methane or ammonia, which remain liquid at temperatures where water would freeze solid. Others consider life that might use silicon instead of carbon as its structural backbone, or biochemistry based on entirely different principles. They also consider life that might exist in extreme conditions, such as environments bathed in high radiation, temperatures near the boiling point of water, or dark caves without any sunlight at all. On Earth, life has shown remarkable adaptability that repeatedly surprised scientists. Organisms called extremophiles live in boiling hot springs where temperatures exceed what was once thought compatible with biology. Others thrive in the crushing pressures and frigid darkness deep under the ocean floor, kilometers below the seabed. Some live inside solid rock, extracting energy from minerals. Others endure the dry, frozen valleys of Antarctica where temperatures remain below freezing for most of the year and where conditions resemble those on Mars. These discoveries show that life can exist in places once thought impossible. They also expand the range of environments we might consider as potential habitats elsewhere. Life appears to need certain fundamentals, but the specific conditions can vary far more widely than early researchers imagined.

Despite decades of searching with increasingly sophisticated instruments, no direct evidence of extraterrestrial life has been found. This absence does not imply that life does not exist elsewhere. It simply reflects the vastness of the universe and the profound difficulty of exploring it. The nearest stars beyond the Sun are trillions of kilometers away. Even the most advanced telescopes can only observe exoplanets indirectly, through the faint effects they have on their parent stars or the thin sliver of starlight that filters through their atmospheres. Searching for life on the moons of Jupiter and Saturn requires missions that take years to reach their destinations and must operate in harsh radiation environments with limited power and communication. Yet each new discovery brings us closer to answering the question. We now know that planets are common rather than rare, with most stars hosting at least one planet. We know that water is widespread throughout the galaxy, found in ice on asteroids, in the atmospheres of gas giants, and likely in oceans beneath the surfaces of distant moons. We know that the ingredients for life, the carbon and other elements forged in stars, are found throughout the cosmos in molecular clouds and on planetary surfaces. These facts suggest that the conditions needed for life are not rare cosmic accidents but natural outcomes of how matter behaves under the laws of physics and chemistry.

The search for intelligent life adds another layer to the investigation, one that reaches beyond microbes or simple organisms. Scientists listen for signals that might indicate communication from distant civilizations. They examine radio waves, which travel well through space and require relatively little energy to produce, as well as optical pulses from lasers, and other forms of energy that could carry deliberate messages across interstellar distances. Large radio telescopes scan the sky systematically, listening for signals that show patterns impossible to create through natural processes, such as narrow frequency ranges or repeating sequences. So far, no confirmed signals have been detected, only natural emissions from pulsars, quasars, and other cosmic phenomena. Several explanations might account for this silence. It is possible that intelligent life is exceedingly rare, arising only under very specific circumstances that Earth happened to meet. It is also possible that intelligent species communicate in ways we don't yet understand, using technologies or methods that our current instruments cannot detect. Another possibility lies in timing. Civilizations may arise and fade over timescales that rarely overlap with one another. The universe is nearly fourteen billion years old, and if technological civilizations typically last for only thousands or millions of years, the chances of two civilizations existing simultaneously and close enough to communicate could be vanishingly small. The window for communication may be narrow both in space and time.

The search for life beyond Earth is ultimately a reflection of our own curiosity and our need to understand our context in the universe. We want to know if life is a common outcome of cosmic processes, emerging readily wherever conditions permit, or a rare event that happened perhaps only once. We want to understand whether the universe is filled with living worlds, each following its own evolutionary path, or whether Earth stands alone, surrounded by lifeless planets under empty skies. This search deepens our understanding of how life begins, requiring us to study the chemistry that bridges non-living matter and biology. It illuminates how life survives by examining the limits of adaptation and the range of conditions organisms can tolerate. It sheds light on how life evolves by asking whether evolution follows similar patterns elsewhere or takes entirely different forms under different circumstances. The search also reminds us that Earth, with all its diversity and complexity, is part of a larger universe that may contain surprises we have not yet imagined. We are made of elements forged in stars, and our existence is connected to cosmic processes that operate throughout space and time.

Whether life exists elsewhere or not, the search expands our knowledge and shapes our perspective in profound ways. It encourages the development of new technologies, from more sensitive telescopes to advanced instruments that can analyze atmospheric chemistry from light years away. It drives new methods of exploration, including spacecraft that can navigate autonomously and instruments that can detect life in forms we might not expect. It inspires new ways of thinking about the universe, challenging us to imagine possibilities beyond our experience while remaining grounded in what science reveals. The search also connects us to the stars in a tangible way by reminding us that the carbon in our cells, the calcium in our bones, and the iron in our blood all began in ancient suns that died long before Earth formed. The search for life beyond Earth continues because it reflects humanity's enduring desire to understand its place in a universe filled with possibilities. It represents one of the deepest questions we can ask, and even the search itself, regardless of what we find, expands the boundaries of human knowledge and imagination.

Chapter Twenty Two
Fate of the Sun

The Sun feels permanent. It rises every morning, warms the planet, powers the weather, and drives the growth of plants. It has shined for billions of years, and for all of human history it has been the most stable presence in the sky. To a human lifetime of decades, or even to the span of recorded civilization spanning mere thousands of years, the Sun appears utterly unchanging. Yet the Sun is not unchanging. It is a star, and like all stars, it has a life cycle shaped by the physics of light, heat, and gravity. The Sun was born, it will age, and one day it will reach an end. Understanding the Sun's fate reveals how fragile Earth's future is, and how closely life on this planet depends on the star that anchors it.

The Sun began its life roughly 4.6 billion years ago as a cloud of gas and dust collapsing under gravity. As this cloud contracted, it heated up, and the center became increasingly hot and dense. When the temperature in the core reached approximately ten million degrees Celsius, a threshold was crossed. At this temperature, hydrogen nuclei moving fast enough could overcome their electromagnetic repulsion and collide with sufficient energy to fuse together. Four hydrogen nuclei combined through a series of reactions to form one helium nucleus, and in the process, a tiny fraction of mass converted into energy according to Einstein's famous equation. This fusion released enormous amounts of energy that created an outward pressure, balancing the inward pull of gravity. This fusion has powered the Sun since its birth and continues to fuel it today.

Over time, however, the Sun has slowly changed. Fusion steadily increases the amount of helium in the core and decreases the amount of hydrogen. Helium nuclei are heavier than hydrogen nuclei, so as helium accumulates, the core becomes denser. To maintain the pressure needed to support the Sun's outer layers against gravity, the core must become slightly hotter. This increased temperature causes fusion to proceed at a faster rate, releasing more energy. As a result, the Sun's luminosity gradually increases. It is brighter now than it was in Earth's early history, roughly thirty percent more luminous than when the planet first formed, and it will continue to brighten as long as it fuses hydrogen into helium. This gradual brightening has already transformed Earth's climate over geological time, though life has adapted through countless changes in atmospheric composition and ocean chemistry.

In about one billion years, the Sun's increasing brightness will make Earth much less hospitable. The additional energy reaching our planet will raise surface temperatures significantly. Higher temperatures will cause the oceans to lose water through evaporation at an accelerating rate. The atmosphere will hold more moisture, and since water vapor is itself a greenhouse gas, it will trap additional heat and accelerate the warming process further. This creates a feedback loop that intensifies over time. As temperatures continue to climb, the rate of chemical weathering will increase, removing carbon dioxide from the atmosphere. Eventually, carbon dioxide levels may drop so low that most plants can no longer perform photosynthesis efficiently. As the oceans shrink, the climate will shift dramatically. Weather patterns will intensify, bringing more violent storms and greater temperature extremes. Continents will heat beyond comfortable limits for most complex life forms. Life as we know it will struggle to survive. Microorganisms adapted to extreme heat might endure in protected environments like deep caves or underground aquifers, but the world will no longer resemble the planet that humans inhabit today. The biosphere that has flourished for billions of years will face its greatest challenge.

As the Sun continues to age, it will eventually exhaust the hydrogen in its core. This will happen in approximately five billion years from now. When the hydrogen fuel runs out, fusion in the core will cease. Without the outward pressure from fusion, gravity will compress the core, squeezing it to higher densities and temperatures. As the core contracts and heats, a shell of hydrogen surrounding the core will become hot enough to begin fusing. This shell burning produces even more energy than core fusion did, and the tremendous energy output pushes the outer layers of the Sun outward. The Sun will grow into a red giant, a phase that will reshape the entire inner solar system. Its radius will expand to perhaps one hundred times its current size or more, extending its outer atmosphere far beyond its present boundary. This expansion will engulf Mercury and Venus, vaporizing both worlds entirely. Earth may also fall victim to this expansion. Current models suggest Earth might be consumed by the Sun's outer layers or at least brought close enough to be scorched until nothing recognizable remains. Even if Earth remains just outside the Sun's physical reach, the intense heat and radiation will strip away the atmosphere molecule by molecule and boil away any remaining water. The planet will be reduced to a barren, molten rock.

The red giant phase won't last forever, perhaps only a few hundred million years. As the core continues to contract under the weight of helium ash, its temperature will climb higher and higher. When the core temperature reaches approximately one hundred million degrees Celsius, helium nuclei will suddenly begin fusing into carbon and oxygen through a process called the helium flash. Once helium in the core begins to fuse into these heavier elements, the Sun will stabilize briefly, entering a phase where it burns helium in its core while continuing to fuse hydrogen in a surrounding shell. But this phase will be short compared to the billions of years of stable hydrogen burning that came before. When the helium is depleted, the Sun will no longer be able to sustain nuclear fusion in its core. The Sun lacks sufficient mass to compress its core to the temperatures needed to fuse carbon and oxygen into heavier elements. Without fusion energy to support it, the outer layers will be expelled. Pulsations will develop as the star becomes unstable, and strong stellar winds will carry the outer envelope away into space, creating a glowing shell of gas known as a planetary nebula. This nebula, despite its name having nothing to do with planets, will be illuminated by ultraviolet radiation from the remaining core, which will collapse into a white dwarf. The Sun will no longer produce energy through fusion. It will simply shine with the residual heat of its former self, gradually radiating away the thermal energy stored in its compressed interior.

A white dwarf is a dense, hot remnant about the size of Earth but containing most of the Sun's original mass, roughly half or more compressed into a sphere no larger than our planet. The matter in a white dwarf is supported not by thermal pressure from fusion but by electron degeneracy pressure, a quantum mechanical effect that prevents electrons from being squeezed into the same state. This exotic state of matter is so dense that a teaspoon of white dwarf material would weigh several tons on Earth. The white dwarf will initially glow with a surface temperature hotter than the Sun is today, shining with a brilliant white or blue-white light. It will glow faintly for many billions of years as it slowly cools, radiating its stored heat into the cold void of space. Over time, it will shift from white to yellow to red as its temperature drops. After tens of billions of years, longer than the current age of the universe, it will become so cool that it no longer emits visible light. It will fade into a black dwarf, a cold, dark cinder floating silently through space. The solar system will continue to exist around it, but it will no longer have the warm, stable light that the Sun provides today. Planets that remain will orbit in perpetual darkness around a star that no longer supports the conditions needed for life. The age of the Sun will have ended.

Despite the dramatic changes that lie ahead, the Sun's life cycle is a normal one. It is a typical star of moderate size, classified as a G-type main sequence star. Stars more massive than the Sun evolve faster and end their lives in more spectacular fashion, sometimes exploding as supernovae. Stars less massive than the Sun burn their fuel more slowly and can shine for trillions of years. The Sun occupies a middle ground, and its evolution follows the same pattern as countless similar stars throughout the galaxy. Understanding the Sun's fate helps us understand the future of Earth and the challenges humanity may face in the far future. The timescales involved are almost incomprehensibly long. A billion years is more than two hundred times longer than the age of complex animal life on Earth. Five billion years is longer than our planet has existed. Yet these timescales are finite and certain. If humans or their descendants survive long enough, they will need to leave Earth long before the Sun reaches its red giant stage. Planets around other stars, carefully selected for their stability and longevity, or artificial habitats built in the depths of space may become the next homes for life. The technology required for such migrations is far beyond our current capabilities, but the distant future may demand it.

The Sun's story also reminds us of the temporary nature of our environment. The stability that supports life is not guaranteed forever. It is the result of a delicate balance between the Sun's energy output, Earth's distance from that star, the composition of our atmosphere, and countless other factors. This balance will eventually shift as the Sun evolves. Earth exists in a window of time where conditions are just right, a period sometimes called the habitable epoch. But that window won't remain open indefinitely. The conditions that allowed life to emerge, evolve, and flourish are exceptional, arising from a particular moment in the life cycle of our star. Knowing this does not diminish the beauty of the world we inhabit. Instead, it deepens our appreciation for the moment in which we live. We exist during an extraordinary period when liquid water covers much of the planet's surface, when the atmosphere is rich with oxygen, and when a star of perfect brightness warms the world without burning it.

The fate of the Sun is a powerful reminder that nothing in the universe remains unchanged. Stars live, shine, and eventually fade. Planets form, evolve, and someday vanish. Life appears, adapts, and seeks new possibilities when old ones are foreclosed. The universe itself continues to change over unimaginable timescales, driven by forces that operate whether or not anyone is present to observe them. Understanding these cycles allows us to see our place within a larger story, one that stretches long before us and will continue long after. We are participants in a cosmic drama that unfolds over billions of years, witnesses to a brief moment of stability in a dynamic and evolving cosmos.

Chapter Twenty Three
The Future of Humanity

Humanity exists in a remarkable moment. We live at a time when the Sun is stable, producing energy at a steady rate that has remained nearly constant for millions of years. The climate is suitable because Earth currently occupies a position within an interglacial period, a temporary warming phase between ice ages that has lasted roughly eleven thousand years. This relatively mild climate has allowed agriculture to flourish and civilizations to develop. Earth supports a wide range of life, with ecosystems that span from the deepest ocean trenches to the highest mountain peaks, each adapted to its environment through billions of years of evolution. Human beings have learned to understand the laws of nature through careful observation and experimentation. We have explored the planets with robotic probes and human missions, mapped their surfaces, and analyzed their atmospheres. We examine the universe with instruments that reach far beyond the limits of our senses, detecting radiation across the electromagnetic spectrum and capturing images of galaxies billions of light years away. Yet this moment is temporary. The Sun won't remain stable forever. Its nuclear reactions will eventually intensify, causing it to grow hotter and larger. Earth won't remain hospitable indefinitely. Changes in solar output, atmospheric composition, and geological activity will eventually make the planet uninhabitable for complex life. Humanity's future depends on how we respond to the challenges that lie ahead, both on our own planet and beyond it.

The first step in understanding humanity's future is recognizing that our species is still young. Modern humans appeared approximately three hundred thousand years ago, a brief interval on the scale of Earth's four and a half billion year history. Civilizations, defined by agriculture, permanent settlements, and written language, have existed for only about ten thousand years, a small fraction of even human existence. For most of human history, survival depended entirely on the immediate environment. People lived close to the land, gathering food from wild plants, hunting animals, and moving with the seasons. Their lives were guided by the cycles of nature, the availability of water, and the patterns of weather. Today, humanity has expanded far beyond those limitations. We have created technology that changes the environment in profound ways, redirecting rivers, clearing forests, and altering the chemical composition of the atmosphere. We explore space with satellites, probes, and crewed missions that venture beyond Earth's protective envelope. We connect people across the globe through networks of communication that transmit information at the speed of light. These abilities give us opportunities to solve problems, share knowledge, and extend our reach into previously inaccessible domains. They also place responsibilities on us that earlier generations never faced. We now possess the power to influence planetary systems, to determine which species survive and which vanish, and to shape the conditions that future generations will inherit.

One of the greatest challenges humanity must confront is the health of the planet. Human activity has altered the atmosphere by releasing carbon dioxide and other greenhouse gases that trap heat and warm the climate. The concentration of carbon dioxide in the atmosphere has increased by more than forty percent since the beginning of the industrial era, rising from about 280 parts per million to over 420 parts per million. This increase has already warmed the planet by approximately one degree Celsius, triggering changes in weather patterns, ocean temperatures, and ice sheet stability. The effects of this warming are not uniform. Some regions experience more intense droughts, others face increased flooding, and many see shifts in the timing of seasons that disrupt agriculture and ecosystems. These changes occur at a rapid pace, faster than many species can adapt through natural selection. Industrial growth, energy production, and expanding populations have increased the pressure on Earth's resources. Forests are cleared for agriculture, oceans are depleted of fish, and freshwater supplies are strained by demand. The decisions we make about energy, agriculture, and conservation will determine the planet's stability for centuries. If humanity chooses to protect the environment and develop sustainable technologies such as renewable energy sources, efficient resource use, and carbon capture systems, Earth can remain a home for future generations. These technologies already exist in various forms, but they require widespread adoption and continued improvement. If we ignore the warning signs visible in rising temperatures, melting ice, and ecosystem collapse, the consequences will affect every part of life. Food production will become less reliable, coastal cities will face flooding from rising seas, and millions of people will be displaced by environmental changes that make their homes uninhabitable.

Another important aspect of the future concerns the growth of technology. Advances in medicine, engineering, and artificial intelligence will continue to reshape society in ways that are difficult to predict but impossible to ignore. Medical technologies such as gene editing, personalized treatments based on individual genetics, and advanced prosthetics that interface directly with the nervous system can extend life, reduce suffering, and improve quality of life for billions of people. Engineering advances in materials science, energy storage, and construction methods can open new opportunities for sustainable development and space exploration. Artificial intelligence, which already assists with everything from medical diagnosis to traffic management, will become more sophisticated and more integrated into daily activities. These technologies can improve efficiency, reduce waste, and solve problems that would overwhelm human capabilities alone. They can also introduce risks if used without caution. Autonomous systems may make decisions that lack human judgment or ethical consideration. Genetic modifications could have unintended consequences that spread through populations. Artificial intelligence might be used to manipulate information, invade privacy, or concentrate power in ways that undermine democracy and individual freedom. Humanity must decide how to integrate these tools into daily life while preserving dignity, safety, and fairness. The choices made today about regulation, education, and access will shape the world of tomorrow, influencing everything from health care to communication to the nature of work. These decisions require wisdom, foresight, and a willingness to prioritize long term wellbeing over short term convenience.

Beyond Earth, new possibilities await. The search for life beyond our planet has expanded dramatically with the discovery of thousands of exoplanets orbiting other stars. Many of these planets exist in the habitable zone where liquid water can exist on their surfaces. The universe contains environments where life might exist, from the subsurface oceans of Europa and Enceladus to the methane lakes of Titan to the atmospheres of distant worlds we have only begun to characterize. If humanity discovers even the simplest organisms elsewhere, perhaps bacteria living beneath the ice of an outer solar system moon or microbes floating in the clouds of Venus, it will transform our understanding of biology and our place in the cosmos. Such a discovery would demonstrate that life is not unique to Earth, that the chemical and physical processes that led to life here can occur elsewhere under different conditions. The presence of life beyond Earth would show that the universe is capable of supporting life in multiple locations, increasing the urgency to explore, learn, and protect those environments. It would also raise profound questions about the diversity of life, the possibility of intelligence elsewhere, and the ethical responsibilities we bear toward other living systems.

Space exploration will play a major role in humanity's long term survival. As the Sun ages over the next several billion years, its core will gradually heat up as hydrogen becomes scarce and helium accumulates. This heating will cause the Sun to brighten, increasing the energy Earth receives. Within approximately one billion years, this increased radiation will boil away Earth's oceans, creating a runaway greenhouse effect that will make the planet uninhabable for life as we know it. Long before that final stage, Earth will become increasingly less stable as temperatures rise and water becomes scarce. Our species will need new places to live if it wishes to continue existing on timescales measured in millions of years. The Moon offers a nearby stepping stone for research and development. Its low gravity makes it easier to launch materials into space, and its surface contains resources such as water ice in permanently shadowed craters and minerals that could support a permanent base. Establishing a presence on the Moon would teach humanity how to live beyond Earth, how to recycle resources in closed systems, and how to protect people from the harsh environment of space. Mars provides a world that, with enough preparation, could one day support human settlements. Its day length is similar to Earth's, its surface contains water ice and the raw materials needed for construction and fuel production, and its atmosphere, though thin, offers some protection from radiation. Terraforming Mars, the process of modifying its atmosphere and temperature to make it more Earth like, would require centuries of sustained effort and technological capabilities we don't yet possess. More distant destinations, such as the icy moons of the outer solar system, which harbor subsurface oceans beneath their frozen crusts, or artificial habitats in space, rotating cylinders or spheres that simulate gravity through spin, may eventually become homes for future generations. These steps will require long periods of effort, innovation, and cooperation among nations and institutions. Space exploration is expensive, dangerous, and slow, but it represents an investment in the survival of our species.

The fate of the Sun adds another dimension to humanity's future. Although the Sun's most dramatic changes lie approximately five billion years in the future, when it will exhaust its hydrogen fuel and expand into a red giant that engulfs the inner planets, they remind us that Earth is not a permanent sanctuary. No planet lasts forever. Every world in the universe has a finite lifetime determined by the evolution of its star, the stability of its orbit, and the geological processes that shape its surface. Human beings will need to expand beyond this planet if they wish to survive for millions of years. This expansion won't happen quickly. Interstellar travel remains beyond our current capabilities, limited by the vast distances between stars and the enormous energy required to accelerate spacecraft to even a small fraction of light speed. It will require sustained commitment across many generations, scientific progress in propulsion, life support, and radiation protection, and the ability to adapt to unfamiliar environments with different gravity, different atmospheres, and different resources than those found on Earth. But the same intelligence that allowed human beings to create societies, develop tools, and build civilizations can guide humanity into new worlds. The capacity for abstract thought, planning, cooperation, and innovation that enabled our ancestors to survive in hostile environments on Earth can be applied to the challenges of living in space. We are not passive observers of our fate. We are active participants with the ability to shape our future through the choices we make.

The future of humanity also depends on how we treat one another. As communication improves through the internet, satellite networks, and translation technologies, and the world becomes more interconnected economically and culturally, conflicts and misunderstandings can spread more easily. A dispute in one region can affect markets and politics across the globe within hours. Misinformation and divisive rhetoric can reach billions of people almost instantly, amplifying tensions and undermining trust. At the same time, cooperation can spread more easily as well. Scientific discoveries are shared across borders. Humanitarian aid reaches disaster zones faster. People from different cultures exchange ideas, art, and knowledge in ways that were impossible for previous generations. The ability to work together across cultures and nations will influence the stability and progress of society. Global challenges such as climate change, pandemic disease, and resource scarcity cannot be solved by individual nations acting alone. They require coordinated action, shared resources, and a willingness to prioritize collective wellbeing over narrow self interest. Humanity's future will be stronger if it is built on cooperation, respect, and shared goals rather than conflict and division. The survival of our species depends not only on our technological capabilities but also on our social and political capacity to work together toward common objectives.

The long term survival of humanity is not guaranteed, but it is possible. We have the knowledge and ability to protect the planet through sustainable practices, clean energy, and conservation efforts that preserve biodiversity and ecosystem function. We can expand into space through incremental steps, building infrastructure on the Moon and Mars, developing closed loop life support systems, and eventually establishing self sufficient settlements beyond Earth. We can continue the journey that began with the first cells on Earth, those simple organisms that appeared nearly four billion years ago and gave rise through evolution to all the complexity and diversity of life we see today. The decisions we make today will shape the world that future generations inherit. Our species has reached a point where its actions influence not just local environments such as forests or rivers, but the entire planet through changes in atmospheric composition, global temperature, and the distribution of species. Eventually, our influence may extend to the solar system itself as we mine asteroids, terraform planets, and build habitats in locations that are currently empty space.

Humanity's future is a continuation of the story that began billions of years ago when the universe formed the first stars from clouds of hydrogen and helium left over from the Big Bang. Those stars forged heavier elements in their cores and released them into space when they exploded as supernovae. Those elements condensed into new stars and planets, including our Sun and Earth. It is the next chapter in a long sequence of events that includes the rise of life, which emerged from chemical reactions in Earth's early oceans, the evolution of intelligence, which developed over hundreds of millions of years as nervous systems became more complex, and the development of consciousness, the capacity for self awareness and abstract thought that allows us to contemplate our own existence and plan for the future. Humanity has the ability to guide this chapter. We can explore new worlds and learn about the universe through observation and experiment. We can protect the environments that sustain us by managing resources wisely and limiting our impact on ecosystems. We can adapt to new challenges by developing technologies and social structures that promote resilience and cooperation. The universe will continue to change according to the laws of physics, which govern everything from the motion of galaxies to the behavior of atoms. Humanity will change with it, evolving biologically over long timescales and culturally over short ones, transforming ourselves and our environments as we have throughout our history.

Chapter Twenty Four
Unanswered Questions

Even with all the knowledge humanity has gained, the universe remains filled with mysteries that we have not yet solved. These mysteries don't signal failure or weakness in science. They reveal the depth of reality and the limits of what we can currently measure or understand. Every discovery opens the door to new questions. Each answer uncovers more layers to explore. What once seemed like a simple question about how stars shine led us to nuclear physics, which in turn revealed the strange behavior of particles at quantum scales. What began as observations of planetary motion evolved into relativity, which transformed our understanding of space and time themselves. The universe is not a completed story. It is an unfolding one, and unanswered questions mark the places where the next chapters will be written.

One of the most profound mysteries concerns the nature of consciousness. We understand the structure of the brain and the way neurons communicate through electrical impulses and chemical signals. We can map the regions responsible for language, memory, visual processing, and motor control. We can study how damage to specific areas affects behavior and perception. We can observe which patterns of neural activity correspond to different mental states. Yet the experience of being aware remains difficult to explain. Consciousness is more than the sum of electrical signals and chemical reactions. It includes the feeling of being alive, the ability to reflect on our own thoughts, and the sense of having a personal identity that persists through time. There is something it is like to see the color red, to taste chocolate, to feel sadness. These subjective experiences, sometimes called qualia, cannot be fully captured by describing brain states alone. A complete map of every neuron and synapse would still leave the central question unanswered. How does the firing of neurons give rise to the inner world of experience? How does matter become aware of itself? This puzzle sits at the boundary between biology, physics, and the nature of existence itself. Some researchers believe consciousness emerges from the complexity of neural networks, while others suggest it may require principles we have not yet discovered. The mystery remains open.

Another major mystery involves dark matter. Observations show that galaxies rotate faster than expected based on the visible matter they contain. When astronomers measure the speed of stars orbiting in the outer regions of galaxies, they find these stars moving too quickly to be held in place by the gravity of visible matter alone. According to the laws of motion and gravity, these stars should fly apart into space. Yet the galaxies remain intact. Something unseen generates the extra gravity that holds galaxies together. This invisible mass does not emit light or interact with ordinary matter in familiar ways. It does not absorb, reflect, or scatter electromagnetic radiation. It reveals itself only through its gravitational influence on visible matter and light. Dark matter makes up most of the matter in the universe, accounting for roughly five times as much mass as ordinary matter. Yet we don't know what it is made of. Scientists have proposed various candidates, including exotic particles that barely interact with normal atoms. Experiments buried deep underground search for these particles, hoping to detect rare collisions. Other experiments look for signs of dark matter annihilating in space or being produced in particle accelerators. Despite decades of effort, no detection has been confirmed. Its true nature remains hidden, a shadow we can see only by its effects.

Dark energy presents an even greater puzzle. It appears to act as a force that pushes the universe outward, causing the expansion of space to accelerate. This discovery came from observations of distant supernovae, which appeared dimmer than expected. Their faintness indicated they were farther away than predicted, suggesting that the expansion of the universe has been speeding up over time rather than slowing down. Dark energy influences the largest scales of reality, shaping the future of the cosmos. It dominates the energy content of the universe, making up roughly seventy percent of everything that exists. Yet scientists don't know what dark energy is, why it exists, or whether its strength changes over time. It may be a property of space itself, a kind of energy woven into the fabric of the vacuum. Some theories connect it to the cosmological constant, a term Einstein once introduced and later called his greatest blunder. Others propose that dark energy might be a dynamic field that varies across space or time. Whatever its origin, dark energy remains one of the most important unanswered questions in cosmology. It will determine whether the universe continues expanding forever, eventually tearing apart even atoms, or whether some other fate awaits.

The beginning of the universe also contains mysteries that science cannot yet fully explain. We can describe what happened after the universe began expanding from an incredibly hot and dense state. We understand how particles formed, how hydrogen and helium were created in the first minutes, and how gravity pulled matter together into the first stars. But we don't know what caused the beginning itself. General relativity predicts that as we trace time backward, we arrive at a point where density and temperature become infinite, a singularity where the equations break down and predictions become meaningless. At this boundary, space and time as we understand them cease to exist. There is no clear answer to what existed before space and time took shape, or whether the concept of "before" has any meaning in that context. Time itself may have emerged along with space, making questions about what came earlier impossible to formulate. Some theories propose that the universe emerged from quantum fluctuations in a pre-existing state, a kind of uncertainty in nothingness that allowed something to appear. Others suggest that our universe may be one bubble in an eternal multiverse, with countless other universes existing beyond our reach. Still others explore the possibility that time is cyclical, with universes being born, expanding, and collapsing in an endless sequence. None of these ideas has been proven. The true origin of the universe remains unknown, perhaps forever beyond the reach of observation.

The possibility of life elsewhere raises questions we have not yet answered. We know that the ingredients for life are common throughout the cosmos. Carbon, hydrogen, oxygen, and nitrogen exist in abundance. Water has been found on moons, asteroids, and planets across the solar system. Planets exist in every corner of the galaxy, orbiting most stars, with many residing in regions where temperatures allow liquid water to exist on their surfaces. But we don't know whether life arises easily or rarely. We don't know whether the chemistry that led to life on Earth occurs naturally wherever conditions are suitable, or whether it requires a rare and improbable sequence of events. We don't know whether living organisms exist on distant moons like Europa or Enceladus, far away planets orbiting other stars, or worlds with conditions unlike anything found on Earth. Life might thrive in environments we consider hostile, beneath ice sheets or in acidic clouds. It might take forms we would not recognize, using different chemistry or operating on different principles. We have only begun exploring the environments where life might exist. Robotic probes have visited a handful of worlds. Telescopes have analyzed the atmospheres of a few distant planets. Until we find clear evidence, whether in the form of microbes on Mars, chemical signatures in an exoplanet's air, or signals from an intelligent civilization, the question remains open. The answer will reshape our understanding of biology and our place in the cosmos.

Time itself raises questions that remain unresolved. We understand how time relates to speed, gravity, and the shape of space. Einstein showed that time is not absolute but flows at different rates depending on motion and gravitational fields. We can measure how time flows more slowly near massive objects and how clocks tick faster at higher altitudes. Satellites must account for these differences to maintain accurate positioning systems. But the deeper nature of time remains mysterious. Why does time flow in one direction rather than another? Most physical laws work equally well whether time moves forward or backward. The equations of motion describe a ball thrown into the air just as accurately when run in reverse. Yet we experience time moving only forward. Eggs break but don't spontaneously reassemble. We remember the past but not the future. This arrow of time appears connected to entropy, the tendency of systems to move from order to disorder. The universe began in a highly ordered state and has been spreading into increasing disorder ever since. But this merely shifts the question. Why did the universe begin in a state that allowed time to move forward toward increasing complexity? Why was the initial condition so special? The answers to these questions remain elusive, pointing toward deep connections between time, thermodynamics, and the origin of the cosmos.

Even the smallest scales of existence contain mysteries. Quantum mechanics describes particle behavior with extraordinary accuracy, making predictions confirmed by countless experiments. Yet the underlying reasons for its strange rules remain unclear. Particles can behave like waves, spreading across space and interfering with themselves. They can exist in multiple states at once, occupying different positions or having different properties simultaneously until measured. This superposition allows particles to explore many possibilities at the same time. When a measurement occurs, the particle settles into a single definite state, but we cannot predict which state it will choose. We can only calculate probabilities. Even more mysteriously, particles can become entangled, forming connections that persist across great distances. Measuring one particle instantly affects its partner, no matter how far apart they are. Einstein called this spooky action at a distance and doubted its reality, yet experiments have confirmed that entanglement is real. These behaviors challenge our understanding of reality, suggesting that the universe at its most fundamental level operates according to rules very different from our everyday experience. The deeper explanation that unites quantum mechanics with gravity, reconciling the physics of the very small with the physics of the very large, has not yet been found. String theory, loop quantum gravity, and other approaches attempt to bridge this gap, but a complete theory remains out of reach.

Unanswered questions don't diminish the value of what we know. They give us direction and purpose. They show where curiosity is needed and where understanding can grow. Science advances not by answering every question but by asking better questions, by refining our understanding and expanding the boundaries of knowledge. The universe is so vast and intricate that no single generation could uncover all its secrets. Some mysteries may require technologies we have not yet invented, mathematical tools we have not yet developed, or ways of thinking we have not yet imagined. Each mystery invites future scientists, explorers, and thinkers to continue the work. These open questions are seeds of discovery, whispering that there is more to learn and more to understand. They remind us that humility is as important as knowledge, that recognizing what we don't know is as valuable as celebrating what we do.

Humanity stands at a point in history where many great discoveries have already been made, yet the largest mysteries remain unsolved. We have decoded the structure of DNA, landed probes on distant worlds, and detected gravitational waves rippling through space. We have looked back in time to see the universe as it was billions of years ago and looked inward to see the quantum machinery underlying reality. Yet for all we have learned, the deepest questions remain unanswered. These mysteries will shape the path of science in the centuries to come. They will drive new experiments, inspire new theories, and challenge new generations to push beyond current limits. They remind us that the universe is deeper than we can imagine and more complex than we have yet realized. The unanswered questions are not obstacles. They are opportunities. They are the spaces where new ideas will grow, new technologies will develop, and the next chapters of human understanding will begin. They are the frontier where wonder meets knowledge, where imagination guides investigation, and where the human story continues its long journey toward comprehension.

Chapter Twenty Five
Why Anything Exists

Among all the questions humanity has ever asked, none reaches deeper than the question of why anything exists at all. The universe could have been empty. It could have contained no stars, no planets, no space, no time, and no laws for anything to follow. Instead, it contains galaxies stretching across unimaginable distances, filled with matter and energy that shaped itself over billions of years. It contains life that arose from simple beginnings and developed the ability to think, feel, and wonder about its own existence. The presence of something rather than nothing is not merely surprising. It is profoundly mysterious, challenging our deepest assumptions about reality itself.

Science has achieved remarkable success in describing how the universe works. It can explain how the universe expanded from an extraordinarily hot, dense state, how matter cooled and organized itself into atoms, and how stars and planets emerged from clouds of gas and dust. These explanations trace events back through cosmic history, following a chain of physical processes that stretches billions of years into the past. They reach all the way to the first fractions of a second after the beginning, when the universe was unimaginably small and energetic. But even these powerful explanations encounter a fundamental limit. They describe how the universe evolved, but they don't explain why it came into being in the first place. The beginning describes a process, a sequence of events unfolding according to natural laws. It does not provide a reason, a purpose, or an underlying necessity that demanded existence rather than emptiness.

The challenge deepens when we consider what it means to ask about the origin of everything. Before space and time existed, our everyday understanding of cause and effect cannot apply in familiar ways. Every cause we observe happens at a particular moment in time and leads to an effect at a later moment. But if time itself began with the universe, then asking what happened before the beginning may be a question without meaning. The concept of "before" depends entirely on the existence of time, and time is not something that existed independently and then contained the universe. Time is part of the universe, woven into its very fabric. This creates a peculiar situation where the origin of existence sits outside the framework we use to understand origins. It is like asking what lies north of the North Pole. The question uses familiar words, but it reaches beyond the domain where those words have clear meaning.

Despite these conceptual challenges, physicists have developed ideas that attempt to address the question of why something exists. One of the most intriguing proposals suggests that the universe may have arisen naturally from the laws of quantum physics. This idea begins with a surprising discovery about the nature of empty space. What we think of as emptiness, as the absence of everything, is not truly empty according to quantum theory. Even in a perfect vacuum, with no particles present and no energy flowing through it, space itself contains a residual energy. This quantum vacuum is not a static, inert nothingness. It is dynamic, filled with fluctuations that arise from the fundamental uncertainty built into quantum mechanics.

These fluctuations are not merely mathematical abstractions. They have measurable physical consequences. In the quantum vacuum, pairs of particles and antiparticles constantly appear and disappear, existing for such brief moments that they remain within the limits set by quantum uncertainty. These are called virtual particles, and their effects have been detected in laboratory experiments. They influence the behavior of atoms, slightly shifting the energy levels of electrons in ways that match theoretical predictions with extraordinary precision. The vacuum, in other words, has structure and activity. It obeys laws, contains energy, and permits processes that would be impossible in a truly empty state.

If the quantum vacuum can produce temporary fluctuations, then under the right conditions, one of these fluctuations might not disappear. Instead of vanishing back into nothingness, it could expand. In the earliest moments of such an expansion, a process called cosmic inflation could take over, driving the growth at an exponential rate. This inflation would stretch a region smaller than an atom to astronomical size in a fraction of a second, creating space, time, matter, and energy all at once. If this scenario is correct, then the universe exists because the laws of quantum physics permit it. The laws themselves would be fundamental, forming the foundation from which space, time, and everything within them emerge. Existence, in this view, is not a miracle that violated natural law. It is a consequence of natural law operating under extraordinary conditions.

This explanation is scientifically compelling, but it immediately raises another question, one that may be even more profound. Where did the laws of quantum physics come from? Why do they exist at all? Why do they take the specific form that allows matter, energy, space, and time to appear? These laws are not arbitrary or random. They are precise, mathematical, and remarkably consistent. They govern the behavior of every particle in the universe, from the smallest quarks inside atomic nuclei to the motion of entire galaxies. They don't change from one location to another or from one moment to the next. This universality suggests a deep underlying structure to reality, a set of principles that holds throughout all of space and time. Yet the origin of these principles remains mysterious. They may be part of an even larger reality that we have not yet discovered, governed by deeper laws that give rise to the physics we observe. Or they may simply exist without any further explanation, as the most basic facts of reality that cannot be reduced to anything more fundamental.

Some physicists and cosmologists have explored the possibility that our universe may not be the only one. This idea, known as the multiverse hypothesis, suggests that a larger reality might contain countless universes, each with different properties and different physical laws. These universes would not interact with each other, and we could never observe them directly. But if this scenario is correct, it would change how we think about existence. In a multiverse, some universes might expand so rapidly that they tear themselves apart before any structure can form. Others might collapse back on themselves almost instantly, vanishing as quickly as they appeared. A few might have laws of physics that prevent atoms from forming, or that make stars impossible, or that allow only the simplest particles to exist. Only a small fraction of all possible universes would have the particular combination of properties that permits complexity, stability, and the emergence of life.

In such a framework, our universe would be one successful outcome among an enormous number of possibilities. The fact that we exist to observe the universe and ask questions about it would not be surprising. We couldn't find ourselves in a universe that didn't allow life to appear. This reasoning is known as the anthropic principle, and it offers a way to explain why the universe seems so finely tuned for our existence without requiring design or purpose. The answer would simply be that we observe the kind of universe capable of producing observers. Yet even this idea, as elegant as it may be, does not resolve the deepest question. It explains why our particular universe has the properties it does by appealing to a larger reality containing many universes. But it does not explain why that larger reality exists. It does not tell us why there is something rather than nothing. It shifts the question to a broader scale without ultimately answering it.

Another approach considers whether the question itself might be shaped by human expectations that don't apply to the origin of everything. We are used to living in a world where events have causes.

When something happens, we naturally look for the reason behind it. This pattern of thinking serves us well in daily life and in scientific investigation. Every discovery in physics, chemistry, and biology follows chains of cause and effect, tracing outcomes back to the conditions that produced them. But the origin of the universe may not follow the same pattern. The beginning of space, time, matter, and the laws that govern them may not have a cause in the way we understand causation. It may simply be the moment when existence began, not preceded by anything and not requiring a prior cause. This possibility is difficult to accept because it contradicts our intuitions, but our intuitions were shaped by experiences within the universe, not by the origin of the universe itself.

A different perspective suggests that existence might actually be simpler and more natural than nonexistence. At first, this seems counterintuitive. Nothingness sounds like the simplest possible state, the default condition from which anything else would be a departure. But in physics, nothingness turns out to be far more complicated than it appears. True nothingness is not just the absence of matter or energy. It must also be the absence of space itself, the absence of time, the absence of any laws governing anything, and the absence of any possibilities. It must have no properties whatsoever, no structure, and no potential for change. Achieving such absolute emptiness might be extraordinarily difficult, perhaps even impossible.

Consider what it would take to maintain perfect nothingness. There could be no quantum fluctuations, because fluctuations require laws that describe them. There could be no space through which anything could move or in which anything could exist. There could be no time in which processes could unfold. The complete absence of everything is so restrictive, so demanding of total emptiness, that it may be less stable or less natural than a universe governed by simple laws. If the most fundamental state of reality permits even the smallest possibility of structure or change, then existence might emerge inevitably. In this view, something does not arise from nothing through a special act or a violation of logic. Rather, absolute nothingness is so difficult to maintain, so contrary to even the most basic possibilities, that existence becomes the path of least resistance.

This way of thinking about the question does not provide a final answer, but it offers a new lens through which to view the problem. It suggests that we may have been asking the question backwards. Instead of wondering why there is something, we might ask why there would ever be nothing. If existence requires only the possibility of structure and change, while nothingness requires the complete and absolute absence of all possibilities, then existence may be the more natural state. The universe might exist not because something caused it, but because the alternative is too restrictive to be realized.

Science may never provide a complete answer to why there is something rather than nothing. The question sits at the boundary between physics and philosophy, where empirical investigation reaches the limits of what can be tested and observed. We can study the behavior of particles, measure the expansion of space, and trace the history of cosmic evolution. But we cannot step outside the universe to examine why it exists in the first place. The tools of science are powerful, but they are tools designed to work within the universe, using its laws and its structure. When we ask about the origin of those laws and that structure, we are asking a question that may exceed the reach of scientific method.

Yet exploring the question remains valuable, even if a final answer proves elusive. The question invites us to think deeply about the nature of reality and the conditions that make our existence possible. It reminds us that the universe is not something we should take for granted. Its existence is not obvious, not inevitable, and not explained by anything more basic that we have yet discovered. The universe is extraordinary simply in the fact that it is here at all. Every galaxy, every star, every planet, every living organism, and every conscious thought depends on the fact that something exists instead of nothing. Without existence, there would be no physical laws, no space for matter to occupy, no time for processes to unfold, and no observers to notice the absence of everything.

The existence of something rather than nothing is the foundation of everything else. It is the reason particles formed in the early universe, the reason stars began to shine, and the reason planets could support the chemistry of life. It allowed evolution to take place across billions of years, complexity to build from simplicity, intelligence to emerge from biological processes, and consciousness to arise with the ability to look back at the cosmos with wonder. If nothing had existed, none of these events would have been possible. There would be no history, no change, no stories to tell, and no minds to contemplate the mystery. The fact that something exists is the precondition for all of science, all of philosophy, and all of human experience.

Chapter Twenty Six
Your Place in the Universe

Standing within the universe, it is easy to feel small. The night sky stretches across the darkness, filled with stars too distant to reach and too numerous to count. Our nearest stellar neighbor, Proxima Centauri, lies more than four light years away, a distance so vast that light itself, traveling at three hundred million meters per second, requires over four years to bridge the gap. The galaxies extend beyond what the eye can see, each containing hundreds of billions of stars, and the observable universe encompasses perhaps two trillion galaxies in total. The universe itself expands far beyond what the mind can grasp, stretching across billions of light years in every direction. Yet the vastness that surrounds us does not diminish our importance. Instead, it reveals the remarkable truth that we exist as part of a universe that allows life, intelligence, and awareness to arise. Meaning does not come from size or scale. It comes from connection, understanding, and the ability to reflect on the story that brought us here.

Everything that makes up your body began in stars that lived and died long before the Earth formed. This is not metaphor or poetry. It is scientific fact, confirmed by our understanding of stellar nucleosynthesis and the periodic table of elements. The carbon that forms your cells, comprising nearly twenty percent of your body mass, was born in ancient stellar cores where temperatures reached tens of millions of degrees. Under these extreme conditions, helium atoms fused together through a process called the triple alpha reaction, creating the carbon that would eventually become part of your DNA, your proteins, and every organic molecule within you. The oxygen you breathe, which makes up sixty five percent of your body, was created in massive stars that exploded as supernovae, releasing their enriched material across space at speeds of thousands of kilometers per second. The calcium in your bones, providing structure and strength, comes from the final moments of dying stars, forged in the intense heat of stellar collapse. The iron in your blood, carrying oxygen to every cell, once existed in the heart of a star heavy enough to collapse under its own weight, creating elements through fusion reactions that required enormous gravitational pressure. Even the trace elements essential to life, the zinc in your enzymes, the copper in your proteins, the iodine in your thyroid, were created in stellar furnaces or scattered by neutron star collisions. You are not separate from the universe. You are built from it, atom by atom. Every element within you carries a history that stretches back billions of years, a lineage that traces through stellar cores, supernova shockwaves, and the slow aggregation of dust clouds that eventually formed our solar system.

The knowledge that we come from the universe gives new meaning to our existence. The universe didn't create us as an afterthought or an accident, but neither were we designed in advance. We emerged from its natural processes, following the laws of physics, chemistry, and biology that govern how matter behaves under different conditions. We are connected to the earliest particles that formed in the initial moments after the Big Bang, to the first molecules that assembled when the universe cooled enough to allow atoms to form, to the first cells that arose when organic chemistry crossed the threshold into self replication, and to the long chain of evolution that followed over nearly four billion years. Our ability to think about this connection, to trace the steps that led from quarks to consciousness, makes humanity unique among the many forms of life on Earth. We are the part of the universe that can look back at itself and ask questions about its origin, its structure, and its purpose. We can contemplate the forces that shaped galaxies and calculate the age of the cosmos. This capacity for self reflection represents something genuinely remarkable within the physical universe. Matter has organized itself in such a way that it can understand its own existence.

Understanding our place in the universe also allows us to appreciate the uniqueness of life and the narrow circumstances under which it can exist. Life is fragile, dependent on conditions that exist only within narrow limits. It requires stable temperatures that allow liquid water to persist, typically between zero and one hundred degrees Celsius, though some organisms can survive in slightly wider ranges. It requires sources of energy, whether from sunlight, chemical gradients, or thermal vents, that can drive the reactions necessary for metabolism. It requires environments that can support growth, providing the elements needed for biological molecules and protecting organisms from harmful radiation. Earth provides these conditions through a fortunate combination of factors. Our distance from the Sun places us within the habitable zone where temperatures permit liquid water. Our atmosphere shields us from cosmic radiation and maintains pressure that keeps water in liquid form. Our magnetic field deflects charged particles from the solar wind. The presence of a large moon stabilizes our axial tilt, preventing extreme climate variations. Yet Earth provides these conditions only for a limited time in cosmic terms. In roughly one billion years, the Sun's increasing luminosity will render Earth too hot for life as we know it. In five billion years, the Sun will expand into a red giant, likely consuming our planet entirely. Knowing this encourages us to treat our planet with care during the brief window in which it can support us. It reminds us that we are temporary visitors in a world shaped by billions of years of natural history, and that the conditions we depend on are neither permanent nor guaranteed.

Meaning also arises from the ability to form relationships, experience emotions, and create communities. These experiences add depth to life in ways that science alone cannot fully describe. They emerge from the complex interactions of neurons, hormones, and consciousness, yet they transcend simple physical explanation to become something subjectively real and personally significant. They are part of what it means to be human. The universe provides the physical laws and materials that support life, establishing the constants and forces that allow chemistry to create complexity. But humans give meaning to their existence through connection and understanding. We interpret our experiences, learn from them, and share them with others. We build languages to communicate our thoughts, develop art to express our emotions, and create knowledge that accumulates across generations. In this way, meaning grows and evolves through time, shaped by cultures, discoveries, and the accumulated wisdom of those who came before us. Each generation inherits the understanding of previous generations and adds its own insights to the collective human experience.

Your place in the universe is shaped by both the vastness around you and the small moments within your own life. The universe is immense, spanning scales that challenge comprehension. A human lifetime represents less than a billionth of the age of the universe. The Earth itself is a pale blue dot, as Carl Sagan observed, suspended in a sunbeam, barely visible when photographed from the outer solar system. Yet it is the small experiences that give life personal value and emotional resonance. A conversation that changes your perspective. A memory that connects you to your past. A feeling of curiosity or wonder when you first understand a new idea. A moment of connection with another person that reminds you of your shared humanity. These experiences matter not because they are large in cosmic terms, but because they reflect the consciousness that the universe has allowed to develop within biological organisms. They are expressions of the same organizational complexity that formed galaxies from hydrogen clouds, stars from collapsing nebulae, and planets from orbiting dust. The processes that created structure in the universe ultimately created minds capable of appreciating that structure.

Perspective comes from realizing that the universe will continue long after our time has passed. The timeline extends in both directions far beyond human comprehension. Stars will continue to be born in molecular clouds, fusing hydrogen into helium and lighting up the darkness of space. They will die in different ways depending on their mass, some gently shedding their outer layers as planetary nebulae, others exploding violently as supernovae, and the most massive collapsing into black holes. Galaxies will move through space, drawn together by gravity or pushed apart by the expansion of space itself. The Milky Way will collide with the Andromeda galaxy in approximately four billion years, a process that will take millions of years to complete. The Sun will expand, consuming Mercury and Venus, and fade as its hydrogen fuel runs out. Eventually it will cool into darkness, becoming a white dwarf that slowly radiates away its remaining heat over trillions of years. Yet for the brief period in which humanity exists, a span of perhaps a few million years within a universe that has existed for nearly fourteen billion years, we have the ability to understand the universe and appreciate its beauty. We can observe distant galaxies and calculate when their light began its journey. We can decode the cosmic microwave background radiation and read the story of the universe's infancy. We can look at rocks and determine their age through radioactive decay. This awareness, this capacity to understand our context within the vast sweep of cosmic history, gives our existence significance even within a timeline that stretches far beyond our own.

Meaning does not need to be assigned from outside the universe. It does not require external validation or supernatural purpose. It grows from our ability to reflect on our existence, imagine futures that don't yet exist, and create things that would not arise through natural processes alone. It emerges from the knowledge that we are part of something larger than ourselves, connected to the history of matter and energy that began with the Big Bang. It comes from the understanding that our thoughts and actions shape the future in ways that extend beyond our individual lives. The knowledge we generate, the stories we tell, the ideas we share, and the changes we make to our world ripple forward through time, affecting those who come after us. The universe sets the stage, providing the physical laws that govern how matter behaves and the initial conditions from which complexity could emerge. Humanity writes its own part of the story, making choices, building civilizations, and seeking to understand the cosmos that created us.

One truth remains clear. You are part of a universe that could have been empty, devoid of structure or complexity, yet it gave rise to matter, then to life, then to intelligence, and ultimately to awareness capable of understanding its own origins. You carry within you the history of stars, encoded in every atom of your being. You carry the legacy of evolution, reflected in your DNA and the biological mechanisms that sustain you. And you carry the ability to understand your own origins, to trace the path from the first moments of the universe to the present moment. In that understanding, you find your place in the universe. And in that place, you find meaning.

Chapter Twenty Seven
What Do You Believe

At the end of this long journey through the universe, one question remains. It is not a question that science can answer with equations or measurements. It is a question that belongs to you alone. After exploring time and space, particles and atoms, stars and planets, life and consciousness, what do you believe about the universe and your place in it? Science can describe how everything works. It can trace the history of the cosmos from its first explosive instant to the present moment and explain the processes that shaped matter into life. It can show how stars formed from collapsing gas clouds, how molecules assembled according to the principles of chemistry, how evolution produced intelligence through natural selection, and how the brain generated awareness through billions of neural connections. But science cannot tell you what these things mean to you. It cannot measure the significance of your existence or determine the purpose of the cosmos. Only you can decide that.

Throughout this book you have walked through a universe guided by natural laws. You have seen how particles arose from energy in the first fractions of a second after the Big Bang, how atoms formed when the universe cooled enough for electrons to bind to nuclei, and how stars forged the heavier elements through nuclear fusion, elements that now exist in your blood and bones. You have followed the steps that led from simple single celled life to complex organisms with specialized organs and intricate systems, from the rise of intelligence in early animals to the emergence of self aware consciousness in the human mind. You have seen how the universe shaped you before you were ever born, how every atom in your body was processed through the core of an ancient star, how the chemistry that makes your thoughts possible arose from the basic laws of physics. You have learned what modern science believes about the origins of everything that exists. You have encountered the evidence collected from telescopes and particle accelerators, from fossil records and genetic sequences, from experiments repeated countless times across laboratories around the world. Now the question becomes what you believe after knowing all of this.

Some readers may feel strengthened in their view that the universe is a place governed entirely by natural processes. They may see the scientific explanation as complete in itself, requiring no additional layer of meaning beyond the elegant workings of physical law. They may find beauty and wonder in the fact that unconscious matter, following simple rules over vast stretches of time, gave rise to complexity, to life, to minds capable of understanding their own origins. For them, the natural world is enough. The story of matter becoming conscious of itself may be all the meaning necessary. Others may feel that the scientific story points toward something deeper. They may believe that the laws of nature reflect an underlying purpose or intention that science cannot measure, that the remarkable consistency of physical constants and the emergence of consciousness suggest more than mere chance. They may see the universe not as an accident but as a creation, with the scientific account describing the method rather than negating the meaning. Some may feel that science and faith can exist side by side without contradiction. Science can explain how the universe unfolded, mapping each step of the journey from hydrogen to human beings, while faith can speak to why it exists at all, addressing questions of purpose that lie outside the reach of empirical investigation. No single answer is required. The universe, in its vastness and complexity, allows space for many perspectives.

Your beliefs may be shaped by science, philosophy, faith, personal experience, or questions that remain unanswered despite centuries of inquiry. You may find meaning in the natural processes that created the world, seeing purpose in the unfolding of cause and effect across cosmic time. Or you may believe that something beyond nature gave rise to it, that the universe itself points to a source that transcends the physical realm. You may feel that consciousness has a deeper source than biology, that subjective experience cannot be fully explained by neural activity alone. You may wonder why the laws of physics are so precise, why the fundamental constants take exactly the values needed to allow stars and planets and life to form. You may question why the universe produces life at all, when it could just as easily have remained a sterile expanse of matter spreading through empty space. These questions belong to you. They are not settled by data or resolved by theory. They ask you to think beyond what science can measure and reflect on what the story of the universe means for your own life, for your sense of purpose, for how you choose to live within this brief window of awareness you have been granted.

Science does not ask you to abandon your beliefs. It does not demand that you accept a universe empty of meaning simply because it has revealed the mechanisms behind natural phenomena. It invites you to understand the universe more clearly, to see past the surface of things and grasp the deeper patterns that govern reality. It shows you how matter forms, how life grows, and how the cosmos evolves according to principles that can be tested and verified. It reveals patterns that have existed since the first moment of time, mathematical relationships that hold true across billions of light years and billions of years. Whether you see these patterns as purely natural, arising from the inherent properties of matter and energy, or as part of something greater, as expressions of a deeper order or intention, is your choice. Science provides the tools to understand the world. It offers you knowledge of how things work. Belief gives you the lens through which you interpret it, the framework through which you assign meaning and significance to what you have learned.

This is not meant to persuade you in any direction. This chapter does not exist to convert you to naturalism or to argue for spirituality. It is meant to give you space to consider your own view, to acknowledge that the scientific account, however complete it may be in describing physical processes, does not dictate what you must believe about ultimate questions. You have walked through the scientific story of the universe, from its birth in unimaginable heat to its current state of stars and galaxies and living worlds. You have seen the evidence that supports it, drawn from observations of distant supernovae and measurements of cosmic background radiation, from laboratory experiments and mathematical models. You have followed the path from the Big Bang to the rise of humanity, understanding each major transition that led from elementary particles to conscious beings. Now you can decide what this story means to you. You can ask whether you believe the scientific explanation is complete or whether there is something more that science cannot yet describe, something that may forever lie beyond its methods and instruments.

The universe is vast and full of unanswered questions. Dark matter remains mysterious. Dark energy pushes space apart for reasons we don't fully understand. The origin of the laws themselves, the reason why these particular laws exist rather than others, remains unexplained. The nature of consciousness, the hard problem of how subjective experience arises from objective matter, continues to challenge researchers. Science continues to explore these mysteries, and your own beliefs may grow and change alongside that exploration. Whether you see the universe as a product of natural laws, as part of a larger purpose, or as something still beyond full understanding, your perspective matters. You are not a passive observer but an active participant in the universe's story. You are a thinking being in a universe that gave rise to thought, a conscious mind in a cosmos that produced consciousness. That alone gives your beliefs weight. You represent the universe becoming aware of itself, looking back across billions of years to understand its own history.

After reading this book, you carry with you the knowledge of how the universe formed and how life emerged within it. You understand the sequence of events that led from formless energy to structured matter, from lifeless chemicals to living cells, from primitive organisms to beings capable of asking why any of this exists. But you also carry your own interpretation, your own sense of meaning, and your own place within this story. Science has told you what it knows. It has given you the facts as best they can be determined through observation and reason. Now the question belongs to you. What do you believe?

Epilogue

In the beginning there was no space, no time, no matter, no energy. There was not even darkness, for darkness requires space to fill. There was not silence, for silence requires time to pass through. There was nothing as we understand it. Then, in a moment so brief that human language barely touches it, the universe began. Physicists measure this first instant in units called Planck time, each one lasting about 0.00000000000000000000000000000000000000000001 seconds. In those first incomprehensible fractions of a second, all the energy that would ever exist in the universe burst into being, and space itself began expanding faster than the speed of light, carrying that energy outward in what we call inflation.

From a hot, dense state smaller than anything we can imagine, far tinier than an atom, far smaller even than the smallest subatomic particle, space expanded and time unfolded. As the universe grew, it cooled. In the first microseconds, temperatures dropped from trillions of degrees to merely billions. In this cooling primordial fire, the first particles burst into existence. Quarks, the fundamental building blocks of matter, condensed from pure energy. Within the first second, these quarks combined into protons and neutrons, forming the nuclei that would become the universe's first atoms. For nearly 380,000 years, the universe remained too hot for electrons to settle into stable orbits around these nuclei. The cosmos was opaque, a glowing fog of plasma where light couldn't travel freely. Then, as temperatures fell below about 3,000 degrees, electrons finally joined the waiting nuclei to form the first complete atoms. Hydrogen and helium filled the young cosmos, with hydrogen making up about three quarters of all matter and helium most of the rest. These primordial gases drifted through expanding space as the universe continued to cool, spreading outward into a darkness lit only by the fading afterglow of creation itself.

For millions of years, this hydrogen and helium filled the universe with almost perfect uniformity. But not quite perfect. Tiny variations in density, differences of less than one part in a hundred thousand, were enough. Gravity, patient and persistent, gathered this matter toward the denser regions. Clouds began to form. These clouds collapsed under their own weight, growing denser and hotter as gravity compressed them. When the cores of these collapsing clouds reached temperatures of about ten million degrees, something remarkable happened. Hydrogen atoms began fusing together, releasing enormous amounts of energy. The first stars ignited, lighting up the darkness perhaps two hundred million years after the Big Bang.

In those stars, the universe learned to build new elements. The cores of stars are nuclear furnaces where lighter elements fuse into heavier ones. Hydrogen fuses into helium. Helium fuses into carbon. Carbon fuses into oxygen. The process continues through neon, magnesium, silicon, and finally iron. These are the heavy elements that would one day form planets, oceans, and living cells. Stars of different sizes created different elements, with the most massive stars forging the full range of elements all the way to iron in their cores. But stars couldn't create anything heavier than iron through normal fusion because those reactions consume energy rather than release it.

When the largest stars exhausted their fuel, they died in spectacular explosions called supernovae, events so energetic that a single dying star could briefly outshine an entire galaxy. In these violent deaths, temperatures soared so high that even the heaviest elements could be forged. Gold, silver, uranium, and every element heavier than iron were born in supernova explosions. These blasts scattered the newly created elements into space, enriching the interstellar gas with the ingredients for future worlds. The universe was becoming chemically diverse.

New generations of stars formed from this enriched material, stars whose composition included all the elements created by previous generations. Around some of these later stars, disks of dust and gas began to swirl. Within these disks, particles collided and stuck together. Dust became pebbles. Pebbles became boulders. Boulders became planetesimals. Planetesimals became planets. This process of accumulation took millions of years, but eventually produced solar systems with rocky inner planets and gas giants farther out, each shaped by the specific conditions in its region of the disk.

One of those planets was Earth. It formed about four and a half billion years ago in orbit around a medium sized star in an ordinary spiral galaxy. Early Earth was hostile, bombarded by asteroids, covered in molten rock, and shrouded in a toxic atmosphere. But as it cooled, water vapor in the atmosphere condensed into rain that fell for thousands of years, filling the first oceans. Those oceans became chemistry laboratories on a planetary scale.

On Earth's surface, in those ancient oceans warmed by volcanic heat and bombarded by lightning and ultraviolet light, chemistry built complexity. Simple molecules dissolved in water encountered each other billions upon billions of times. Some combinations were stable. Some could catalyze reactions in other molecules. Simple molecules became chains. Chains became more complex structures that could store information in their molecular bonds. Eventually, perhaps in warm tidal pools or near deep sea volcanic vents, molecules appeared that could make copies of themselves using raw materials from their surroundings. This was not life as we know it, not yet, but it was the threshold. When these self replicating systems began competing for resources and passing on variations to their copies, evolution began. Life had started.

Over billions of years, evolution shaped life into forms of staggering diversity. Single cells learned to capture energy from sunlight, releasing oxygen as waste and transforming Earth's atmosphere. Other cells learned to use that oxygen for more efficient energy production. Some cells merged with others in partnerships that became permanent, creating the complex cells that would eventually build all plants, animals, and fungi. Life organized itself into tissues, tissues into organs, and organs into bodies capable of movement and sensation. Nervous systems evolved, first as simple nets of cells that could sense and respond, then as clusters of nerves that could process information, and finally as brains capable of memory, learning, and prediction.

From early animals with primitive nervous systems came intelligence, the ability to solve problems, recognize patterns, and adapt behavior based on experience. From intelligence came self awareness, the recognition of oneself as a distinct entity separate from the environment, an observer capable of reflecting on its own existence. From self awareness came language, art, science, and the ability to ask questions about the universe that created us. Consciousness emerged not as something separate from matter but as what matter does when organized in certain complex ways.

Everything in this book is part of that same continuous chain. There are no gaps, no breaks in the sequence. From the first fraction of a second after the Big Bang to the rise of galaxies to the formation of stars and planets to the appearance of life and the development of consciousness, each step built upon the previous one. The laws of physics shaped chemistry. Chemistry shaped biology. Biology shaped minds. And minds became capable of understanding the laws of physics that started the whole process. The universe built the foundations that eventually produced you, using nothing but time, natural processes, and the fundamental forces embedded in the fabric of reality from the beginning.

FROM THE BIG BANG TO YOU

Your body is made of atoms born in ancient stars. The calcium in your bones, the iron in your blood, the oxygen you breathe were all forged in stellar furnaces and scattered into space by dying stars billions of years ago. Your thoughts are patterns built from chemistry that began in the early oceans, molecular signals passed between neurons using mechanisms refined over hundreds of millions of years of evolution. Your awareness is the latest step in a process that began nearly fourteen billion years ago, a process that started with the simplest possible beginning and built, layer by layer, toward greater complexity.

This is what science believes about the unfolding of reality. It describes the processes, the laws, and the history that shaped the universe and everything within it. It traces cause and effect across billions of years, explaining how the simple became complex, how the uniform became diverse, how lifeless matter became living, and how life became conscious. Science does this through observation, measurement, testing, and reasoning. It builds models that must match what we actually see when we look at the universe. It does not attempt to answer spiritual questions. It does not speak for faith or purpose. Those belong to your own beliefs, to the personal meaning you find in existence.

And in my belief, God is the first scientist and the last scientist. The One who understood the laws before they existed, who set the constants that determine how particles interact, how stars burn, how atoms bond, and how life emerges. The One who sees the end of the story before the first moment ever began, who knows how the long chain of cause and effect will unfold across all of time. Science explains the method, the how of existence. Faith speaks to meaning, to the why. The two can stand side by side without conflict, each addressing different questions, each valid in its own domain.

You are the continuation of the universe's story. You are what happens when matter takes the long path through stars, through biology, through evolution, through history, and becomes capable of understanding itself. You are the universe examining itself, asking questions about itself, marveling at itself. Every atom in your body has a history stretching back to the beginning of time.

This is your place in that story. Not separate from the universe. Not outside it watching from some removed position. But made of it, shaped by it, and connected to all of it from the beginning. You are not just in the universe. You are part of the universe, as much a natural expression of its laws and processes as stars and galaxies are.

And as all of existence continues its long arc into the future, the part you choose to play, your beliefs, your actions, your questions, shapes what will happen next. The universe continues to evolve, and conscious beings like you are now part of that evolution. The future is not yet written.

And just as the universe began with a single moment that set everything in motion, a moment of such profound consequence that it created all of space and time itself, your own journey begins with the simple choice to ask what it all means. That question, emerging from matter that has learned to question itself, is perhaps the most remarkable development in the entire history of the cosmos.

Bonus Story

Under the Night Sky

Toby was twelve and convinced that nothing in the world could beat summer nights in his grandpa's backyard. The grass warmed his back like a living thing, storing the day's heat and releasing it slowly into the cooling air. The sky stretched wide above him, impossibly vast, and the whole world felt quieter in a way that made the stars seem louder. It was the kind of quiet that didn't feel empty but full, as if silence itself had gravity and texture. Fireflies drifted up from the grass and blinked lazily along the fence like they were trying to copy the constellations, their cold green light pulsing in patterns that seemed almost deliberate. Grandpa leaned back in his old lawn chair with a heavy sigh that sounded half dramatic and half real, the metal frame creaking under him in its usual pattern. He folded his hands over the soft rise of his stomach and closed his eyes for just a moment.

"Front row seats to the universe," Grandpa said, opening his eyes to take it all in.

"You say that every time," Toby replied, but there was no complaint in his voice.

"And it's just as true every time."

They watched as the sky deepened from blue to indigo and finally into full night, like someone was slowly turning down a dimmer switch on the sun's lingering glow. One star winked on, bright and alone. Then three more joined it. Then dozens, each one taking its time as if making a grand entrance at a cosmic theater. Toby loved this part of the night most, the slow reveal, like the sky had a secret and was letting him in on it little by little. Each new point of light felt like a gift, a small door opening to something vast, ancient, and timeless.

A cricket chirped near his ear, so close he could hear the individual pulses of its song. Grandpa swatted the air with halfhearted annoyance, his hand moving like it weighed ten pounds. "They always come out when I do," he said.

"They must like you," Toby said.

"They probably think I'm part of the furniture by now. Been sitting in this chair so long I might be growing roots."

Toby laughed and nudged Grandpa's foot with his own, a small gesture that said more than words could.

A few minutes passed while the darkness settled around them like a blanket. The temperature dropped degree by degree, and Toby could feel the shift, the way the air grew sharper and cleaner. He pointed at a bright light low in the west, hanging just above the neighbor's roofline. "Is that a star?"

"That's Venus," Grandpa said. "A planet. It shines very bright this time of year because it's covered in thick clouds that reflect sunlight like a mirror. Brightest thing in the sky after the moon."

"It looks alive," Toby said softly, watching how it seemed to pulse with a smooth, unwavering glow that was different from the twinkling stars around it.

Grandpa smiled at that, the kind of smile that said he liked hearing it more than he expected. "It does, doesn't it? That's because planets don't twinkle like stars. They're close enough that their light comes through steadier."

As they watched, a white streak cut across the sky in a brilliant arc. Toby jerked upward, his whole body tensing with excitement. "A shooting star!"

"Show off came early tonight," Grandpa said, pretending he had ordered it himself with a wave of his hand.

Toby stared at the fading trail, his eyes wide. "It was so fast."

"Most beautiful things are," Grandpa said quietly. "That's a meteor, really. Just a piece of rock or dust burning up as it hits our atmosphere. Traveling maybe forty miles every second, faster than you can blink. All that speed turns it into light for just a moment, and then it's gone."

"Forty miles every second?" Toby whispered, trying to imagine it.

"Give or take. Speed like that makes friction, and friction makes heat, and heat makes that glow we see. The whole thing only lasts a few seconds, but people have been wishing on them for thousands of years."

Toby lay back down and folded his hands under his head, letting the coolness of the grass press against his shoulders. The blades tickled the back of his neck. He watched the stars twinkle like little needles of light poking through a dark curtain, each one dancing in its own way. After a moment he said, "Why do they do that? Twinkle, I mean."

"Because the air between us and them keeps moving," Grandpa said. "Earth's atmosphere is like an ocean of air, always shifting and swirling. Hot patches, cold patches, winds moving in different directions. When starlight comes down through all that motion, it gets bent and wobbled. The light shakes a little as it comes through. Makes them look like they're dancing."

"Is that why someone made the Twinkle Twinkle Little Star song?"

"Probably," Grandpa said. "It's hard to look at a sky full of diamonds and not want to sing about it. That twinkling is just physics, but it feels like magic. Feels like the stars are trying to talk to us."

They were quiet for a few breaths. Toby's mind wandered the way it always did when the sky got big enough to make him wonder if he was even real, if his thoughts and his body were just tiny arrangements of the same atoms that made up those distant suns. Then he said, "Grandpa? Do you think there are people on other planets?"

Grandpa tapped his fingers against the armrest, thinking. The pace was slow and thoughtful. "The universe is really big," he said. "So big that it's hard to imagine we're the only ones in it. There are more stars out there than grains of sand on every beach on Earth. And most of those stars probably have planets of their own. Odds are, somewhere out there, life found a way. Maybe out there somewhere there are people asking the same question you just asked. Maybe their grandpas are sitting in lawn chairs too, looking up at their own sky."

"What if they're looking back at us right now?" Toby asked, his voice filled with wonder.

Grandpa laughed, a warm rumble in his chest. "Then I hope they're seeing my good side. Though from that far away, I'm just a speck on a speck on a speck. We all are."

Toby laughed, but then his tone changed. He hesitated, pulling in a breath that felt heavier than the others. "Grandpa?"

"Yeah, buddy?"

Toby kept his eyes on the stars, letting the distance make it easier to ask. "Is Grandma up there somewhere?"

Grandpa was still for a long moment. The crickets seemed to quiet. Even the breeze eased up, as if the whole backyard was listening. When he finally spoke, his voice was softer than before, gentler in a way that made Toby's chest ache.

"She's not a star," he said. "She's not living on one, and she didn't become one. But I do think she's somewhere beyond what we can see. Somewhere safe. Somewhere peaceful. Somewhere good."

Toby swallowed hard and blinked at the sky. His eyes stung. "So she's not here anymore."

"She's here in a different way," Grandpa said, and he touched his chest, right over his heart. "In the parts of us that remember her. In the things she taught us. In the way you laugh sometimes and sound just like her. Love doesn't disappear, Toby. It doesn't fade or burn out. It just changes where it lives. Energy can't be destroyed, just transformed. That's physics. But it's also true for love."

Toby felt his throat tighten. He scooted a little closer, and Grandpa placed a warm hand on the back of Toby's head for a moment, enough to let him know he wasn't alone in what he felt. His palm was rough and familiar.

They sat like that for a while. The stars shimmered overhead, still and enduring, their light traveling for years and years just to reach this moment. Toby finally whispered, "Do you think she can see us?"

"I think she sees the things that matter," Grandpa said. "And you matter. You always will."

Toby let out a breath he didn't know he had been holding. He looked up again, and the stars seemed a little gentler somehow, a little less distant.

After a quiet minute he said, "Grandpa? Did God make all this? Or was it really science?"

Grandpa scratched his chin thoughtfully, the stubble making a soft scratching sound. "Some people say it has to be one or the other. But that's like saying a painting is either the paint or the painter. Doesn't make much sense when you think about it."

Toby blinked up at him, waiting.

"I think God made the laws," Grandpa continued. "Set the rules. Set the constants that hold everything together. Gravity, light, time, all of it. All the stuff science studies and measures and tries to understand. Science tells us how the universe works. Faith tells us why it's here. So if you ask me if it's God or science, I say yes."

"So both?" Toby asked.

"Both," Grandpa said firmly. "God is the first scientist and the last scientist. Knows how it all works from the inside out. Made the atoms and the forces and then let them dance together. And here we are, part of that dance, trying to figure out the steps."

Toby settled deeper into the grass, feeling the earth solid beneath him. The words felt peaceful. Simple. Right. Like a door opening instead of closing.

A few minutes passed, and another small streak crossed the sky. This one was fainter, barely visible, but Toby caught it in the corner of his vision.

"Another one," he said quietly.

"They like you tonight," Grandpa said. "The Perseids are coming early this year. Whole swarm of dust and ice from an old comet. Earth passes through it every August."

Toby smiled up at the sky, his earlier sadness lifting. "Does the universe know we're looking at it?"

Grandpa laughed again, "Maybe it does. But I know this. When you look up like that, with wonder in your heart, you're part of it paying attention to itself. You're made of stardust, literally. The iron in your blood, the calcium in your bones, all of it was forged in stars that exploded billions of years ago. So when you look up at the stars, you're looking at family. You're looking at home."

Toby thought about that. He thought about Grandma and whether energy really could change forms the way Grandpa said. He thought about planets with people who might be staring back across impossible distances. He thought about the stars twinkling as if they were alive, each one a furnace of light and heat that had been burning for eons. And he felt something warm settle inside him, something that felt a little like understanding and a little like hope and a little like belonging.

Grandpa stood slowly, using his chair for balance. His knees popped, loud enough to make Toby smile. "You ready to go in?"

"Not yet," Toby whispered. "Just a little longer."

Grandpa watched him for a moment, seeing something in his grandson's face that made him nod with approval. "Take your time. The universe is patient. Been here for billions of years. It can wait a few more minutes."

They stayed until the fireflies faded and the night grew cooler, until Toby's fingers started to feel numb and his breath became visible in small puffs. When they finally walked toward the house, their shoes whispering through the grass, Toby looked up one last time. The stars were still there, constant and calm, scattered across the darkness like a promise.

Grandpa looked up too and said softly, like a benediction, "Remember this. The universe is big, but it isn't bigger than your curiosity. As long as you keep wondering, as long as you keep asking questions and looking up at the night sky with open eyes, you'll always belong in it. You'll always have a place in it."

Toby nodded, feeling the truth of it settle into his bones. Tomorrow the sun would rise, and the stars would fade. But they'd be back tomorrow night, and the night after that, waiting. And he would be here, wondering, belonging, part of the vast and beautiful dance that started long before he was born and would continue long after.

Acknowledgments

Writing this book has been a journey through science, curiosity, and the long story of existence, and I am grateful to everyone who encouraged me along the way. I want to thank those who supported this project with conversations, questions, and patience as the chapters grew one by one. I am grateful for the scientists, researchers, and educators whose work made it possible to understand the universe in the first place. Their discoveries form the foundation on which every page of this book is built.

Much of what appears in these chapters also comes from what I have learned over the years. These are ideas I have studied, thought about, and written down across the last couple of decades, slowly collecting pieces of understanding that eventually shaped this book.

To my wife, thank you for your support and for giving me the time I needed to focus, research, and write. To the readers who open this book with curiosity and an open mind, thank you for walking through this journey with me. And to God, the One I believe to be the first scientist and the last scientist, thank you for creating the laws that make this universe discoverable.

Every idea in these chapters comes from a long chain of knowledge built by countless people across many generations. I am grateful to all of them, known and unknown, whose work made it possible for this book to exist.

References

The scientific information presented in this book is based on the current consensus across cosmology, astrophysics, particle physics, planetary science, evolutionary biology, neuroscience, and related fields. The following institutions, research programs, and scientific resources form the core foundation for the explanations and descriptions throughout this book:

Major Scientific Institutions

- NASA, Goddard Space Flight Center
- NASA Jet Propulsion Laboratory (JPL)
- European Space Agency (ESA)
- National Science Foundation (NSF)
- Max Planck Institute for Astrophysics
- CERN and the Large Hadron Collider (LHC)
- Kavli Institute for Cosmology
- SETI Institute

Space Telescopes and Missions

- Hubble Space Telescope research archives
- James Webb Space Telescope (JWST) data releases
- Chandra X-ray Observatory
- Kepler and TESS exoplanet mission archives

- Voyager and New Horizons mission data
- Mars Science Laboratory and Mars rovers (Curiosity, Perseverance)

Scientific Journals and Literature

- Nature
- Science
- Astrophysical Journal
- Monthly Notices of the Royal Astronomical Society
- Physical Review Letters
- Journal of Cosmology and Astroparticle Physics
- Annual Review of Astronomy and Astrophysics
- Annual Review of Neuroscience

Standard Textbooks and Educational Sources

These represent the commonly used foundations of modern scientific education and were used as baseline consensus for the content in this book:

- *Introduction to Cosmology* – Barbara Ryden
- *Cosmology* – Steven Weinberg
- *Astrophysics for Physicists* – Arnab Rai Choudhuri
- *Stellar Structure and Evolution* – Kippenhahn & Weigert
- *Evolutionary Biology* – Douglas Futuyma
- *Molecular Biology of the Cell* – Alberts et al.

• *Neuroscience* – Purves et al.

These sources reflect the consensus framework from which the explanations in this book are derived. They are not cited for specific quotations but as the scientific foundation behind the narrative.

SOURCES CONSULTED

This book draws broadly from decades of accumulated scientific knowledge, including:

Physics & Cosmology

- Big Bang cosmology
- Inflation theory
- Standard Model of particle physics
- General relativity
- Quantum mechanics
- Dark matter and dark energy research

Astronomy & Astrophysics

- Stellar formation models
- Supernova nucleosynthesis
- Planetary formation theories
- Exoplanet detection and classification

Earth Science & Biology

- Origin-of-life research
- Evolutionary biology
- Fossil record data
- Genetics and molecular biology

Neuroscience & Psychology

- Neural network structure
- Consciousness research
- Cognitive evolution

FURTHER READING

Cosmology and the Universe

- *A Brief History of Time* – Stephen Hawking
- *The Fabric of the Cosmos* – Brian Greene
- *Astrophysics for People in a Hurry* – Neil deGrasse Tyson
- *Cosmos* – Carl Sagan

Particle Physics & Quantum Science

- *The Elegant Universe* – Brian Greene
- *The Infinity Puzzle* – Frank Close
- *Quantum* – Manjit Kumar

Life, Evolution, and Biology

- *The Selfish Gene* – Richard Dawkins
- *Your Inner Fish* – Neil Shubin
- *Life: A Universe Within* – John Gribbin

Human Brain & Consciousness

- *How the Mind Works* – Steven Pinker
- *The Feeling of Life Itself* – Christof Koch
- *The Human Brain Book* – Rita Carter

Planetary Science & Exoplanets

- *The Planet Factory* – Elizabeth Tasker

JOEY RAINES

- *The Search for Life on Other Planets* – Bruce Jakosky

NOTES

This book presents scientific concepts in narrative form, summarizing modern consensus rather than cataloging individual studies. Because the information presented is based on thousands of peer-reviewed papers, mission datasets, and decades of research across multiple disciplines, individual citations are not listed for every fact or number.

The explanations in this manuscript represent:

- the standard cosmological model
- mainstream astrophysical theory
- accepted principles of particle physics
- consensus evolutionary biology
- verified neuroscience
- observational data from NASA, ESA, and partner missions

Where uncertainties exist (dark matter, dark energy, origin of consciousness, beginning of the universe), they are noted as such.

Don't miss out!

Visit the website below and you can sign up to receive emails whenever Joey Raines publishes a new book. There's no charge and no obligation.

https://books2read.com/r/B-A-IKQCE-XAEFI

BOOKS 2 READ

Connecting independent readers to independent writers.

Did you love *From the Big Bang to You*? Then you should read *The Loop*[1] by Joey Raines!

[2]

Luke's life looks ordinary from the outside. A stable job. A quiet marriage. Days that blur together without much meaning. But under the surface, something feels off. Too off. He can't shake the sense that he's been here before. Lived this before.

As the pressure builds, Luke starts questioning everything. His choices. His memories. The strange silence in his own soul.

Punishment: The Loop is a haunting, thought-provoking short novel about hidden guilt, second chances, and the quiet pull of something greater calling you to wake up before it's too late.

---

1. https://books2read.com/u/mvwvvX
2. https://books2read.com/u/mvwvvX

Also by Joey Raines

Punishment
The Loop

Standalone
From the Big Bang to You

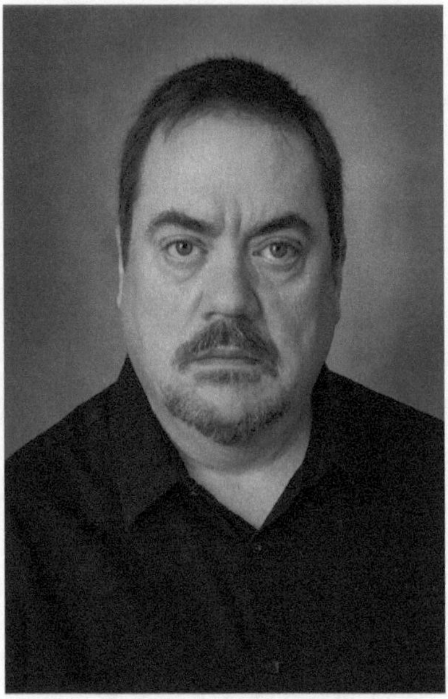

About the Author

Joey Raines is a writer and novelist who blends reality with imagination, truth with fiction, and faith with fear. His work often explores unseen worlds, personal redemption, and the thin line between what we see and what we deny. Joey writes both fiction and nonfiction, drawing from true events, raw emotion, and his belief in something greater than ourselves. Whether crafting a novel or telling a real-life story, his goal is usually the same: to leave readers thinking long after the final page.

MIRROR LOOP
PUBLISHING

About the Publisher

We're an independent publisher committed to bringing you bold, memorable books across diverse genres. Our mission is to deliver stories that resonate long after the final page is turned.

Read more at https://www.mirrorlooppublishing.com.

www.ingramcontent.com/pod-product-compliance
Lightning Source LLC
LaVergne TN
LVHW100304210926
848480LV00003B/1015